WORKBOOK TO ACCOMPANY
RESIDENTIAL CONSTRUCTION ACADEMY

FACILITIES MAINTENANCE

SECOND EDITION

KEVIN STANDIFORD

DELMAR
CENGAGE Learning™

Australia • Brazil • Japan • Korea • Mexico • Singapore • Spain • United Kingdom • United States

DELMAR
CENGAGE Learning™

Workbook to Accompany Residential Construction Academy: Facilities Maintenance, Second Edition
Kevin Standiford

Vice President, Career and Professional Editorial: Dave Garza

Director of Learning Solutions: Sandy Clark

Acquisitions Editor: Jame Devoe

Managing Editor: Larry Main

Vice President, Career and Professional Marketing: Jennifer Baker

Marketing Director: Deborah Yarnell

Marketing Coordinator: Mark Pierro

Production Director: Wendy Troeger

Production Manager: Mark Bernard

Senior Art Director: Casey Kirchmayer

Technology Project Manager: Christopher Catalina

Production Technology Analyst: Thomas Stover

ISBN-13: 978-1-4390-5777-3

ISBN-10: 1-4390-5777-X

Delmar
5 Maxwell Drive
Clifton Park, NY 12065-2919
USA

Cengage Learning is a leading provider of customized learning solutions with office locations around the globe, including Singapore, the United Kingdom, Australia, Mexico, Brazil, and Japan. Locate your local office at:
international.cengage.com/region

Cengage Learning products are represented in Canada by Nelson Education, Ltd.

To learn more about Delmar, visit **www.cengage.com/delmar**

Purchase any of our products at your local college store or at our preferred online store **www.cengagebrain.com**

Notice to the Reader
Publisher does not warrant or guarantee any of the products described herein or perform any independent analysis in connection with any of the product information contained herein. Publisher does not assume, and expressly disclaims, any obligation to obtain and include information other than that provided to it by the manufacturer. The reader is expressly warned to consider and adopt all safety precautions that might be indicated by the activities described herein and to avoid all potential hazards. By following the instructions contained herein, the reader willingly assumes all risks in connection with such instructions. The publisher makes no representations or warranties of any kind, including but not limited to, the warranties of fitness for particular purpose or merchantability, nor are any such representations implied with respect to the material set forth herein, and the publisher takes no responsibility with respect to such material. The publisher shall not be liable for any special, consequential, or exemplary damages resulting, in whole or part, from the readers' use of, or reliance upon, this material.

Printed in the United States of America
2 3 4 5 6 7 8 14 13 12 11 10

Table of Contents

CHAPTER 4 Fasteners, Tools, and Equipment 25

CHAPTER 5 Practical Electrical Theory 37

CHAPTER **8** Surface Treatments 121

CHAPTER **9** Plumbing 129

CHAPTER 17 Blueprint Reading for the Facility Maintenance Technician

Preface

Introduction

Designed to accompany *Residential Construction Academy: Facilities Maintenance, 2nd edition,* this workbook is an extension of *Facilities Maintenance, 2e* that provides additional review questions and exercises designed to challenge and reinforce the student's comprehension of the content presented in the core text.

About the Text

The workbook is divided into chapters, with each chapter directly corresponding to a chapter in *Residential Construction Academy: Facilities Maintenance, 2nd edition.* Each chapter consists of an introduction, objectives, review questions, and job sheets and exercises. The review questions are composed of a variety of matching, true/false, multiple choice, short answer, and essay-style questions based on the materials presented in the core text and workbook.

Job Sheets

Each job sheet consists of an objective for that job sheet, instructions, and either an activity or a checklist. The job sheets range in complexity from entry level to more complex problems that require the student to perform calculations.

Features of This Workbook

- Additional review questions and exercises for *Residential Construction Academy: Facilities Maintenance, 2nd edition*.
- Job sheets with additional exercises and activities designed to reinforce the material presented in the core textbook.

Chapter 1   Customer Service Skills

OBJECTIVES

Upon completion of this chapter, you will be able to:

Knowledge-Based

- List the attributes of great service.
- Identify personal strategies for connecting with the people you are providing service to.
- Explain the importance of understanding the needs of the people you are providing services to and their expectations.

Keywords

Confidence

Competence

Appreciation

Self-talk

Empathy

Honesty

Reliability

Courtesy

Introduction

As stated in the textbook, customer service is a critical aspect of business, and failure to provide good customer service can be detrimental to any business regardless of the nature of that business. This is true in a service-oriented business in which the business is not providing a tangible product.

The key to good customer service is to develop a customer service plan and have it available to all employees before any contact is made with a customer.

A good customer service plan should outline the following:

- Customer service mission statement
- Customer service objectives
- Customer service standards
- Customer service quality assurance

The mission statement should outline the purpose of the customer service portion of a business. For example, "It is the mission of the customer service department to provide our customers with a world-class customer service department."

The objectives should outline what the customer service portion of a business is planning to accomplish. For example:

To provide world-class customer service, it is our goal to:

Answer the phone in a pleasant and courteous manner.
Answer the phone on the first three rings or have the phone forward to a recording.

The customer service standards are those standards that should be used to achieve the objectives outlined. For example, a customer service might include a section on how to correctly answer the phone.

Customer service quality assurance should outline how the company is going to assess the department that will deliver the customer service plan. In addition, this assurance should address how the company will evaluate the effectiveness of the customer service plan and make modifications if necessary.

Chapter Review Questions and Exercises

SHORT ANSWER

1. List the elements that make up attitude.

_____

_____

_____

2. What is the difference between competence and confidence?

_____

_____

_____

3. What is the purpose of a customer service plan?

_____

_____

_____

4. Why is it important to list the customer service objectives on a customer service plan?

_____

_____

_____

5. What is the purpose of a quality assurance section of a customer service plan?

_____

_____

_____

Name: _____   Date: _____

Job Sheet 1: Customer Service

• Create a customer service plan for a small HVAC service company. The company is planning to service makes and models of packaged units as well as install new packaged Trane units. The service company will employ a sales force, repair technicians, and installation technicians.

Instructor's Response

Name: _____   Date: _____

Job Sheet 2: Customer Service

- Create a customer service plan for a small plumbing company. The company will service and install both residential and commercial plumbing systems. The service company will employ a sales force, repair technicians, and installation technicians.

Instructor's Response

Name: _____   Date: _____

Job Sheet 3: Customer Service

- As a facility maintenance technician for an apartment complex, you will be required to interact with the management as well as the tenants. Create a customer service plan that addresses the customer service needs of the apartment complex's tenants as well as the management.

Instructor's Response

Chapter 2

Methods of Organizing, Troubleshooting, and Problem Solving

OBJECTIVES

Upon completion of this chapter, you will be able to:

Skill-Based

⊗ Establish priority of work tasks.

⊗ Assign tasks.

⊗ Carry out work order systems.

⊗ Using the steps outlined in the text to properly trouble a technical issue.

Keywords

Tasks	Assigning tasks	Diagnostics
Priority	Troubleshooting	

Introduction

To successfully manage a project regardless of its size, it must first be broken up into phases and/ or tasks. For example, if a facility manager is given the project of replacing the dishwasher in an apartment complex, then that manager could break up the project as follows:

1. Select a new dishwasher.

2. Transport the new dishwasher to the job site.

3. Remove the existing dishwasher.

4. Install a new dishwasher.

5. Test the new dishwasher.

6. Clean up job site.

7. Dispose the old dishwasher.

Once a project has been broken up into phases, the task associated with each phase can be prioritized and organized in a way in which productivity can be increased.

9

Chapter Review Questions and Exercises

TRUE or FALSE

1. _____ Once a group of tasks have been prioritized, their order cannot be changed.
2. _____ Some workers are overwhelmed with instructions at first and may need to listen to the information repeated more than once.
3. _____ To assign a task to an individual, you simply tell that person what to do.

SHORT ANSWER

4. What is the advantage of planning and organizing a group of tasks before starting?

_____

_____

_____

5. What information should be present on a work-order progress report?

_____

_____

_____

6. List the steps for troubleshooting a problem.

_____

_____

_____

7. What are the steps for solving a technical problem?

_____

_____

_____

8. What should you consider before you assign work to a worker?

_____

_____

_____

9. What should you do to ensure that you are effectively communicating with a worker?

_____

_____

_____

10. What information is necessary for planning for a following week?

_____

_____

_____

Name: _____   Date: _____

Job Sheet 1: Managing a Project

- You are the facility maintenance manager of a large apartment complex. As the manager, you are required to assign projects and tasks to the facility maintenance personnel who report to you. You have been given the project of replacing the electric stoves in several units. As a manager, your first responsibility is to develop a plan for replacing the stoves. This will require you to break up the project into manageable tasks, set a priority for each task, estimate the amount of time needed to complete each task, and assign the appropriate personnel.

Task/Phase	Estimated Time to Complete	Priority	Assigned to	Actual Time to Complete

Instructor's Response

Name: _____ Date: _____

Job Sheet 2: Estimating Priority

- In the previous job sheet, you were asked to divide the project of replacing kitchen stoves into manageable phases/tasks and estimate the amount of time to complete each task. Use the phases/tasks defined in the previous job sheet to assign a priority for each task.

Task/Phase	Estimated Time to Complete	Priority	Assigned to	Actual Time to Complete

Instructor's Response

Name: _____   Date: _____

Job Sheet 3: Assigning a Task

- In the previous job sheet, you were asked to assign the priority of the tasks generated in Job Sheet 1. In this job sheet, you will assign the task to one or more of the key personnel listed below. In addition to assigning the key personnel, you will need to provide justification for your selection. For example, the reason I selected John to install the electric stove is because he has prior experience working with electrical appliances.

Task/Phase	Estimated Time to Complete	Priority	Assigned to	Actual Time to Complete

Personnel

Personnel	Experience
John Smith	Has previous experience working with gas appliances and sheet metal
Kevin Adams	Has previous experience working with landscaping
Jim White	Has previous experience working with electrical appliances
Elizabeth Black	Is a new employee and has no experience in maintenance
J. P. Henderson	Has previous experience working with large-scale industrial and commercial maintenance projects

Justification for Selection

Instructor's Response

Chapter 3   Applied Safety Rules

OBJECTIVES

Upon completion of this chapter, you will be able to:

Knowledge-Based

⊗ Explain the purpose of OSHA.

⊗ Explain the basic safety guidelines and rules for general workplace safety.

⊗ Explain the basic safety guidelines and rules for working with and around an electrical power tool and circuit.

Skill-Based

⊗ Create a basic fall protection plan.

⊗ Work safely with ladders and extension ladders.

⊗ Correctly identify and select the proper fire extinguisher for a particular application.

Keywords

Class A fire extinguishers	Asphyxiation
Class B fire extinguishers	Cardiopulmonary resuscitation (CPR)
Class C fire extinguishers	Frostnip
Class D fire extinguishers	Frostbite
Ground Fault Circuit Interrupter (GFCI)	Personal protective equipment (PPE)
Occupational Safety and Health Administration (OSHA)	

Introduction

Completing an assigned task on time and on budget is a primary concern of every facility manager. It should also be the primary concern of everyone associated with a project. However, the most important aspect of any project is safety. Completing a project on time is pointless if doing so produces an injury or fatality. To help ensure that safety is a prime concern, the federal government established the Occupational Safety and Health Administration (OSHA). OSHA was established within the Department of Labor and was authorized to regulate health and safety conditions for all employers with few exceptions.

Personal protection equipment, also known as PPE, is defined as any equipment, and/or clothing that can be used to protect against the elements of the weather as well as one or more risk associated with health and/or safety. Just as occupations and job requirements and expectations vary from one employer to another, so does the type of PPE necessary to adequately protect employees from a potential danger.

Personal protection is one of the most violated regulations and company policies. It is extremely important that workers properly use, wear, store, and maintain their PPE. Failure to do so can result in costly OSHA fines, property damage, and/or loss as well as injury or even death. When on a job site, it is extremely important for all employees to conduct themselves in a professional manner. Most accidents on a job site are the result of carelessness. Therefore it is critical that all employees be aware of their surroundings at all times and to evaluate the immediate area for possible safety violations and/or hazards.

Chapter Review Questions and Exercises

TRUE/FALSE

1. _____ The primary concern of all facility managers should be to complete an assigned task on time and on budget.

SHORT ANSWER

2. What is the purpose of OSHA?

_____

_____

_____

3. What is the purpose of a fall protection plan?

_____

_____

_____

4. What is a fall arrest system?

_____

_____

_____

5. What are the two categories of ladders?

_____

_____

_____

6. Define the following terms:
   Asphyxiation
   Cardiopulmonary resuscitation (CPR)
   Frostnip
   Frostbite

7. What should you do in the event of a cut?

8. What should you do in the case of exposure?

9. What should you do in the event of electrical shock?

10. What are the Department of Energy recommendations for storing chemicals?

Name: _____   Date: _____

Job Sheet 1: Fall Protection Plan

• You are the facility maintenance manager of a high-rise complex. As the manager, you are required to develop a fall protection plan for a group of window washers. The facility is a thirty-five-story building (385 feet) in which the façade is mostly glass. The window-washing crew consists of three people using a scaffolding platform connected to the roof via trusses and cables.

Instructor's Response

Name: _____   Date: _____

Job Sheet 2: Fire Protection Equipment

• For the following types of fires, identify the correct type of fire extinguisher.

Fire	Fire Extinguisher Type
Office Paper Fire	
Office Wood Fire	
Computer Room Fire	
Industrial Fire (Metal)	
Electrical Fire	
Gas Fire	
Grease Fire	
Plastic Fire	

Instructor's Response

Name: _____   Date: _____

Job Sheet 3: Developing a First-Aid Kit

Reading a Scale

Upon completion of this job sheet, you should be able to use the Internet to research the components necessary to construct an effective first-aid kit.

Procedure

Using the space provided below and the Internet, list the components necessary to construct an effective first-aid kit.

Instructor's Response

Chapter 4 — Fasteners, Tools, and Equipment

Upon completion of this chapter, you will be able to:

Knowledge-Based

- Describe the safe use of tools, including power tools used by facility maintenance technicians.
- Describe, select, and install the proper anchors, fasteners, and adhesives necessary for a specific project.
- Select and properly use the appropriate hand tool for a specific project.
- Select and properly use the appropriate power or stationary tool for a specific project.

Keywords

Box nail

Finishing nail

Duplex nail

Brad

Power tool

Screws

Caulk

Introduction

As a facility maintenance technician, you will be required to perform light repairs and maintenance on build-ing(s) and equipment. Before the repairs can be effectively made, you will need to have the correct tools available and in good working order. In addition, as you make repairs to the facility and equipment, you will have to work with different types of fasteners. As with the use of tools, using the correct fasteners is critical.

Fasteners

Just as important as lumber is the way in which it is fastened and anchored. A *fastener* is a mechanical device used to mechanically join two or more mating surfaces or objects. Fasteners are now on the market for just about any job. They can be used where wood meets wood, concrete, or brick, and most are approved by the Uniform Building Code requirements. However, you should always consult your local building code before selecting a particular type of fastener.

Nails

The following is a list of the various types of nails:

- **Common nail**—most often used of all nails. Used for most applications where special features of other nail types are not needed.
- **Box nails**—used for boxes and crates.
- **Finishing nails**—can be driven below the surface of the wood and concealed with putty so that they are completely hidden.

- **Casing nails**—used for installing exterior doors and windows.
- **Duplex nails**—used for temporary structures, such as locally built scaffolds.
- **Roofing nails**—used for installing asphalt and fiberglass roofing shingles.
- **Masonry nails**—used when nailing into concrete or masonry.

Screws

Screws are used when stronger joining power is needed, or for when other materials must be fastened to wood. The screw is tapered to help draw the wood together as the screw is inserted. Screw heads are usually flat, oval, or round, and each has a specific purpose for final seating and appearance.

Types of Screws

- **Drywall screws**—used to attach drywall to wall studs.
- **Sheet metal screws**—used to fasten metal to wood, metal, plastic, or other materials. Sheet metal screws are threaded completely from the point to the head, and the threads are sharper than those of wood screws. Machine screws are for joining metal parts, such as hinges to metal door jambs.
- **Particleboard and deck screws**—corrosion-resistant screws used for installing deck materials and particleboard.
- **Lag screws**—used for heavy holding and are driven in with a wrench rather than a screwdriver.

Tip: Screw length should penetrate two-thirds of the combined thickness of the materials being joined. Use galvanized or other rust-resistant screws where rust could be a problem.

Screw head shapes are usually determined by the screw types and the pitch and depth of the threads. The most common are oval, pan, bugle, flat, round, and hex. There are also different types of slots for these screws.

Bolts

Nuts and bolts are usually used at the same time: The bolt is inserted through a hole drilled in each item to be fastened together, and then the nut is threaded onto the bolt from the other side and tightened, to give a strong connection. Using nuts and bolts also allows for the disassembly of parts.

Types of Bolts

- **Cap screws**—available with hex heads, slotted heads, Phillips head, and Allen drive.
- **Stove bolts**—either round or flat heads and threaded all the way to the head; used to join sheet metal parts.
- **Carriage bolts**—a round-headed bolt for timber; threaded along part of the shank; inserted into holes already drilled.

Tips for using epoxy:

- One gallon of epoxy will cover 12.8 square feet at 1/8-inch thickness, and 6.4 square feet at 1/4 inch thickness.
- Although epoxy generally is able to withstand high temperatures for short periods, we do not recommend using it in conditions above 200°F.
- To achieve maximum adhesion, remove oil, grease, dirt, rust, paint, and water. Use a degreasing solvent (alcohol or acetone) to remove oil and grease. Sand or wipe away paint, dirt, or rust. Roughing up the surface increases surface area for a better bond.
- Epoxy cures quickly enough that there is significant strength after 3 hours. At 70°F the working time is 15 minutes; however, it is still possible to reposition work after up to 45 minutes.
- Uncured epoxy will clean up with soap and water, or denatured alcohol. Wash contaminated clothing. Cured epoxy can be removed by scraping, cutting, or removing in layers with a good paint remover.

Contact Cement

Contact cement is used to bond veneers or to bond plastic laminates to wood for table tops and counters. Coat both surfaces thinly and allow to dry somewhat before bonding. Align the surfaces perfectly before pressing together, as this adhesive does not allow repositioning. Use in a well-ventilated area.

Types of Fasteners

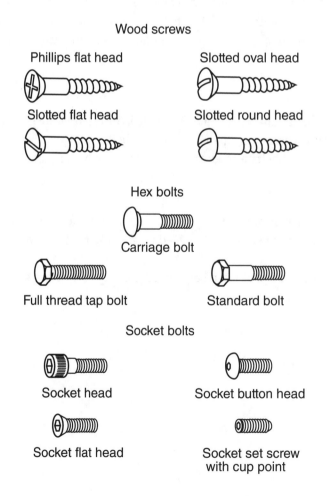

Wood screws

Phillips flat head

Slotted oval head

Slotted flat head

Slotted round head

Hex bolts

Carriage bolt

Full thread tap bolt

Standard bolt

Socket bolts

Socket head

Socket button head

Socket flat head

Socket set screw with cup point

Washers

 Flat washer
(USS and SAE)

 Lock washer

 Lock washer
external tooth

 Lock washer
internal tooth

 Finishing washer

 Dock washer

Sheet metal screws

 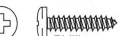 Phillips flat head

 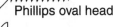 Phillips oval head

  Phillips truss head

  Phillips pan head

  Phillips pan head self drilling

  Hex washer head

 Hex washer head self drilling

Hex washer head self drilling
with sealing washer

Machine screws

 Phillips flat head

 Phillips pan head

  Slotted flat head

 Phillips oval head

  Slotted oval head

  Combination round head

  Combination truss head

 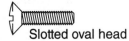 Slotted round head

 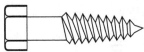 Lag bolt

Nuts

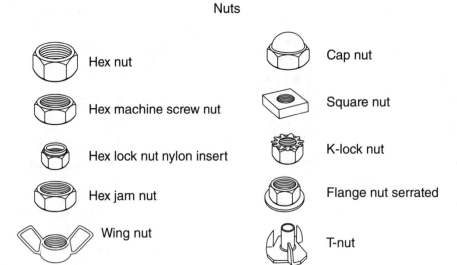

Hex nut

Cap nut

Hex machine screw nut

Square nut

Hex lock nut nylon insert

K-lock nut

Hex jam nut

Flange nut serrated

Wing nut

T-nut

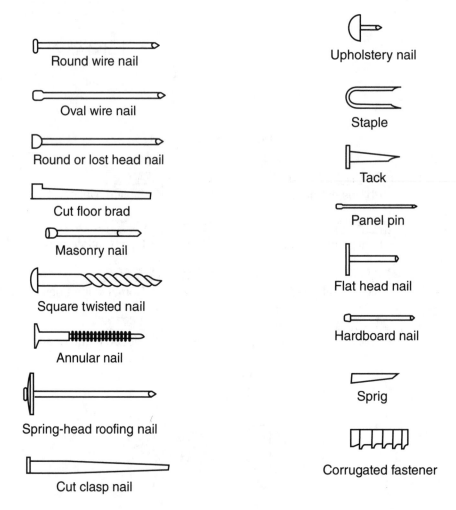

Round wire nail

Upholstery nail

Oval wire nail

Staple

Round or lost head nail

Tack

Cut floor brad

Panel pin

Masonry nail

Flat head nail

Square twisted nail

Hardboard nail

Annular nail

Sprig

Spring-head roofing nail

Corrugated fastener

Cut clasp nail

Chapter Review Questions and Exercises

COMPLETION

1. A _____ is a mechanical device that is used to mechanically join two or more mating surfaces or objects.

2. _____ are used when stronger joining power is needed, or when other materials must be fastened to wood.

3. _____ shapes are usually determined by the screw types and the pitch and/or depth of the threads.

4. _____ is a white, creamy glue, usually available in convenient projects.

5. _____ is the only adhesive with a strength greater than the material it bonds.

6. _____ is used to bond veneers or plastic laminates to wood for table tops and counters.

7. _____ is used as an adhesive during the installation of sinks, countertops, and so on.

8. _____ is used in hundreds of products to improve cleaning efficiency.

9. _____ gallon(s) of epoxy will cover 12.8 square feet at 1/8-in. thickness or 6.4 square feet at ¼-in. thickness.

10. What are plumb bobs used for?

_____

_____

_____

11. How do crosscut saw dirr from rip saws?

_____

_____

_____

12. What are coping saws used for?

_____

_____

_____

13. How do duplex nails different from common nails?

_____

_____

_____

14. What are mason's hammers used for?

_____

_____

_____

15. _____ are available in lengths ranging from 6 to 100 feet.

Name: _____   Date: _____

Job Sheet 1: Selecting the Correct Nail

- Use the chart below to describe the intended use for the nail or staple listed.

Nail Type	Description/Use
Box	
Brad	
Casing	
Common	
Cut Flooring	
Drywall	
Duplex	
Finishing	
Gutter Spikes	
Masonry or Concrete	
Panel	
Roofing	
Sinker	
Spike	
Spiral	
Tack	
Upholstery	

Instructor's Response

Name: _____   Date: _____

Job Sheet 2: Selecting the Right Hammer

- List the intended use for the hammers listed below.

Common nail hammer with curved claw

_____

Rip hammer with straight claw

_____

Finishing hammer

_____

Ball peen hammer

_____

Hand drilling hammer

_____

Soft-face hammer

_____

Tack hammer

_____

Brick hammer

_____

Drywall (wallboard) hammer

_____

Carpenter's mallet

_____

Instructor's Response

Chapter 5 Practical Electrical Theory

OBJECTIVES

Upon completion of this chapter, you will be able to:

Knowledge-Based

- ✪ Understand the basic principle of basic electricity.
- ✪ Describe the difference between AC and DC currents.
- ✪ Understand properties of common electrical wires used by facilities maintenance technicians and understand and correctly measure wire size and load-carrying capacity.
- ✪ Understand the operation and functions of emergency circuits.
- ✪ Describe different types of emergency backup electrical power systems.

Skill-Based

- ✪ Calculate electrical load by using Ohms law.

Keywords

Atom	Electron theory
Matter	Conventional current flow theory
Element	Voltage
Law of charges	Watt
Law of centrifugal force	Alternating current
Valence shell	Direct current
Coulomb	Ohm's law
Ampere (amp)	Continuous load

Introduction

Having a good understanding of electrical theory is essential to being able to effectively work with and on electrical systems. Not understanding the relationship between resistance, voltage, and current can make working with electricity difficult and dangerous.

Series Circuits

Switches and controls are commonly wired in series with each other to control one or more loads. The simplest and easiest electric circuit to understand is the series circuit. The series circuit allows only one path of current flow through the circuit. In other words, the path of a series circuit must pass through each device in the entire circuit. All devices are connected end to end within a series circuit. Figure 5-1 shows a series circuit with four resistance heaters.

Characteristics of a Series Circuit and Calculations for Current, Resistance, and Voltage

The current draw in a series circuit is the same throughout the entire circuit because there is only one path for the current to follow. The current in a series circuit is shown in the following equation:

$$I_t = I_1 = I_2 = I_3 = I_4 = \cdots$$

(The centered dots [$\cdots$] indicate that the equation continues in the same manner until all the elements of that particular circuit have been accounted for.)

The total resistance R_t in a series circuit is the sum of all the resistances in the circuit. The resistance of a series circuit is shown in the following equation:

$$R_t = R_1 + R_2 + R_3 + R_4 + \cdots$$

The voltage in a series circuit is completely used by all the loads in the circuits. The loads of the series circuit must share the voltage that is being delivered to the circuit. Thus, the voltage will be split by the loads in the circuit.

The voltage of a series circuit changes through each load. This change is called the **voltage drop**. The voltage drop is the amount of voltage (electrical pressure) used or lost through any load or conductor in the process of moving the current (electron flow) through that part of the circuit. The voltage drop of any part of a series circuit is proportional to the resistance in that part of the circuit. The sum of the voltage drops of a series circuit is equal to the voltage being applied to the circuit. This is shown in the following equation:

$$E_t = E_1 + E_2 + E_3 + E_4 + \cdots$$

Ohm's law can be used for calculating any part of a series circuit or the total circuit. Figure 5-1 shows a series circuit with four resistance heaters of different ohm ratings. The calculations for the total resistance, the amperage, and the voltage drop across each heater will be calculated by using the circuit shown in Figure 5-1.

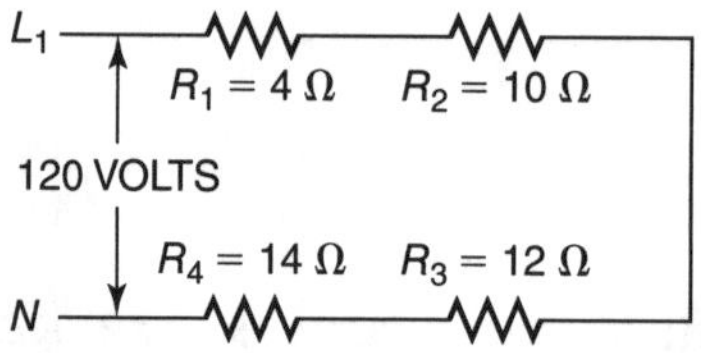

FIGURE 5-1: Series circuit containing four resistance heaters with different resistance values.

The total resistance can be calculated by adding the ohm rating of each heater.

1. Use the formula $R_t = R_1 + R_2 + R_3 + R_4$
2. Substitute the values given in the figure into the formula:

   $R_t = 4 \text{ ohms} + 10 \text{ ohms} + 12 \text{ ohms} + 14 \text{ ohms}$

3. Solve the formula: $R_t = 40 \text{ ohms}$

We use Ohm's law to calculate the amperage draw (current) of the circuit.

1. Use the formula

$$I = \frac{E}{R}$$

2. Substitute the given values into the formula:

$$I = \frac{120}{40}$$

3. Solve the formula: $I = 3$ amperes

Now we use Ohm's law to calculate the voltage drop across each heater.

1. Use the formula $E = IR$ for each resistance
2. Substitute the given values into the formula: $E_{d1} = 3 \times 4$
   (The symbol E_{d1} means the voltage drop across resistance 1.)
3. Solve the formula: $E_{d1} = 12$ volts
4. Solve for each resistance by using the same procedures that were used in steps 2 and 3.

   $E_{d2} = 3 \times 10 = 30 \text{ volts}$ $E_{d3} = 3 \times 12 = 36 \text{ volts}$ $E_{d4} = 3 \times 14 = 42 \text{ volts}$

Note that the total voltage E_t supplied to the circuit is equal to the sum of the voltage drops.

Parallel Circuits

The parallel circuit has more than one path for the electron flow. That is, in a parallel circuit, the electrons can follow two or more paths at the same time. Electric devices (loads) are arranged in the circuit so that each is connected to both supply voltage conductors.

Parallel circuits are commonly used in residential and commercial building because most loads used operate from line voltage. Line voltage is the voltage supplied to the equipment from the main power source of a structure and typically has a value of 115 volts or 240 volts. The parallel circuit allows the same voltage to be applied to all the electric loads connected in parallel, as indicated in Figure 5-2. Note that each load in the circuit is supplied by the line voltage of 115 volts.

Characteristics of a Parallel Circuit and Calculations for Current, Resistance, and Voltage

There will be few occasions when field technicians are required to make calculations for a parallel circuit. This is usually done by the designer of the equipment. However, field technicians should be familiar with the basic concepts and rules of parallel circuits.

The current draw in a parallel circuit is determined for each part of the circuit, depending on the resistance of that portion of the circuit. The total current draw of the entire parallel circuit is the sum of the currents in

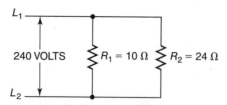

FIGURE 5-2: Parallel circuit for Example 1.

the individual sections of the parallel circuit. The current in each individual circuit can be calculated by using Ohm's law when the resistance and voltage are known. The total ampere draw of a parallel circuit is calculated in the following equation:

$$I_t = I_1 + I_2 + I_3 + I_4 + \cdots$$

The resistance of a parallel circuit gets smaller as more resistances are added to the circuit. The total resistance of a parallel circuit cannot be obtained by taking the sum of all the resistances. It is calculated by the following formula if two resistances are used:

$$R_t = \frac{R_1 \times R_2}{R_1 + R_2}$$

If three or more resistances are located in the circuit, the reciprocal of the total resistance is the sum of the reciprocals of all the resistances (the reciprocal of a number is 1 divided by that number). The following formula is used to calculate the resistance of a parallel circuit with more than two resistances:

$$\frac{1}{R_t} = \frac{1}{R_1} + \frac{1}{R_2} + \frac{1}{R_3} + \frac{1}{R_4} + \cdots$$

The voltage drop in a parallel circuit is the line voltage being supplied to the load. In other words, in a parallel circuit each load uses the total voltage being supplied to the load. For example, if 115 volts are supplied to a load, it will use the total 115 volts. The voltage being applied to each of the four components in Figure 5-2 is the same and is given by the following equation:

$$E_t = E_1 = E_2 = E_3 = E_4$$

Ohm's law can be used to calculate voltage, amperage, or resistance if the other two values are known. You can use Ohm's law to determine almost any condition in a parallel circuit, but pay careful attention to the individual sections of each complete circuit.

Example 1

What is the total current draw of the parallel circuit shown in Figure 5-2?

Solution

1. First calculate the current draw for each individual circuit by using Ohm's law in the form

$$I = \frac{E}{R}$$

2. For I_1 substitute the given values for E and R_1 in the formula and solve:

$$I_1 = \frac{240}{24} = 10 \text{ amperes}$$

3. For I_2 substitute the given values of that circuit in the formula and solve:

$$I_2 = \frac{240}{24} = 10 \text{ amperes}$$

4. Use the formula $I_t = I_1 + I_2$.

5. Substitute in the formula for I_t and solve:

$I_t = I_1 + I_2 = 24$ amperes $+ 10$ amperes $= 34$ amperes

Example 2

Find the total resistance of the parallel circuit in Figure 5-2.

Solution

1. Use the formula:

$$R_t = \frac{R_1 \times R_2}{R_1 + R_2}$$

2. Substitute the known values in the formula:

$$R_t = \frac{10 \times 24 = 7.06}{10 + 24} \text{ ohms}$$

Example 3

What is the resistance of a parallel circuit with resistances of 3 ohms, 6 ohms, and 12 ohms?

Solution

1. Use the formula:

$$\frac{1}{R_t} = \frac{1}{R_1} + \frac{1}{R_2} + \frac{1}{R_3}$$

2. Substitute the known values in the formula:

$$\frac{1}{R_t} = \frac{1}{3} + \frac{1}{6} + \frac{1}{12}$$

3. Mathematical computation of this formula is sometimes difficult. If you have problems, consult your instructor.

$$\frac{4}{R_t} = \frac{4}{12} + \frac{2}{12} + \frac{1}{12} = \frac{7}{12}$$

$7R_t = 12$

$R_t = 1.71$ ohms

Chapter Review Questions and Exercises

SHORT ANSWER

1. What is Ohm's law?

_____

_____

2. What is resistance?

_____

_____

3. What is voltage?

_____

_____

4. What is current?

_____

_____

5. How is electricity produced?

_____

_____

Name: _____   Date: _____

Job Sheet 1: Working with Ohm's Law

Use Ohm's law to make the following calculations.

Resistance	Voltage	Current
100 ohms	5 volts	
1,000 ohms	10 volts	
10 ohms	0.5 volts	
	120 volts	10 amps
	60 volts	2 amps
100 ohms		0.5 amps
	75 volts	5 amps

Instructor's Response

Name: _____   Date: _____

Job Sheet 2: Working with Series and Parallel Circuits

Answer the following review exercises.

True/False

1. _____ The current draw of a parallel circuit is the sum of the current of each branch circuit.

Multiple Choice

2. If three circuits are connected in parallel with a power supply of 30 volts, what would be the voltage supplied to each circuit?
   A. 10 volts
   B. 90 volts
   C. 30 volts
   D. none of the above

3. If two 115-volt loads were connected in parallel with 240 volts, they would: _____
   A. burn dimly.
   B. immediately burn out.
   C. burn correctly.
   D. none of the above.

Short Answer

4. Why are parallel circuits used in the air-conditioning industry?

   _____

   _____

5. What is a series-parallel circuit?

   _____

   _____

6. Why are series-parallel control circuits important in the circuitry used in air-conditioning equipment?

   _____

   _____

7. What is the resistance of a parallel circuit with resistances of 2 ohms, 4 ohms, 6 ohms, and 10 ohms?

   _____

   _____

8. What is the resistance of a parallel circuit with resistances of 10 ohms and 20 ohms?

_____

_____

9. What is the total amp draw of a parallel circuit with ampere readings of 2 A, 7 A, and 12 A?

_____

_____

10. What is the voltage of a series circuit with four voltage drops of 30 volts?

_____

_____

11. What is the resistance of a series circuit with resistances of 4 ohms, 8 ohms, 12 ohms, and 22 ohms?

_____

_____

Instructor's Response

Name: _____   Date: _____

Job Sheet 3: Determining Voltage, Resistance, and Current in a Series Circuit

- Upon completion of this job sheet, you should be able to determine the voltage, current, and resistance in a series circuit.

Complete the following:

Resistance = _____
Voltage = 120 volts
Current = 20 amps

Instructor's Response

Name: _____   Date: _____

Job Sheet 4: Determining Voltage, Resistance, and Current in a Series Circuit

- Upon completion of this job sheet, you should be able to determine the voltage, current, and resistance in a series circuit.

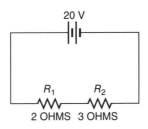

Find the following:

Rtotal = _____ ohms

Itotal = _____ amps

IR1 = _____ amps

IR2 = _____ amps

VR1 = _____ volts

VR2 = _____ volts

Ptotal = _____ watts

PR1 = _____ watts

PR2 = _____ watts

Instructor's Response

Name: _____   Date: _____

Job Sheet 5: Determining Voltage, Resistance, and Current in a Parallel Circuit

- Upon completion of this job sheet, you should be able to determine the voltage, current, and resistance in a parallel circuit.

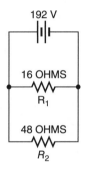

Find the following:

    Rtotal = _____ ohms

    Itotal = _____ amps

      IR1 = _____ amps

      IR2 = _____ amps

     VR1 = _____ volts

     VR2 = _____ volts

  Ptotal = _____ watts

     PR1 = _____ watts

     PR2 = _____ watts

Instructor's Response

Name: _____   Date: _____

Job Sheet 6: Determining Voltage, Resistance, and Current in a Parallel Circuit

- Upon completion of this job sheet, you should be able to determine the voltage, current, and resistance in a parallel circuit.

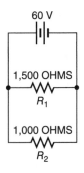

Find the following:

Rtotal =	_____	ohms
Itotal =	_____	amps
IR1 =	_____	amps
IR2 =	_____	amps
VR1 =	_____	volts
VR2 =	_____	volts
Ptotal =	_____	watts
PR1 =	_____	watts
PR2 =	_____	watts

Instructor's Response

Name: _____   Date: _____

Job Sheet 7: Determining Voltage, Resistance, and Current in a Series-Parallel Circuit

- Upon completion of this job sheet, you should be able to determine the voltage, current, and resistance in a series-parallel circuit.

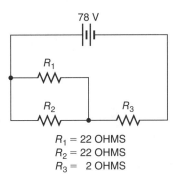

R_1 = 22 OHMS
R_2 = 22 OHMS
R_3 = 2 OHMS

Find the following:

Rtotal = _____ ohms

Itotal = _____ amps

IR1 = _____ amps

IR2 = _____ amps

IR3 = _____ amps

VR1 = _____ volts

VR2 = _____ volts

VR3 = _____ volts

Ptotal = _____ watts

PR1 = _____ watts

PR2 = _____ watts

PR3 = _____ watts

Instructor's Response

Name: _____   Date: _____

Job Sheet 8: Determining Voltage, Resistance, and Current in a Series-Parallel Circuit

- Upon completion of this job sheet, you should be able to determine the voltage, current, and resistance in a series-parallel circuit.

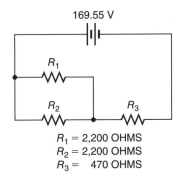

$$R_1 = 2,200 \text{ OHMS}$$
$$R_2 = 2,200 \text{ OHMS}$$
$$R_3 = 470 \text{ OHMS}$$

Find the following:

Rtotal = _____ ohms

Itotal = _____ amps

IR1 = _____ amps

IR2 = _____ amps

IR3 = _____ amps

VR1 = _____ volts

VR2 = _____ volts

VR3 = _____ volts

Ptotal = _____ watts

PR1 = _____ watts

PR2 = _____ watts

PR3 = _____ watts

Instructor's Response

Name: _____   Date: _____

Job Sheet 9: Determining Voltage, Resistance, and Current in a Series-Parallel Circuit

- Upon completion of this job sheet, you should be able to determine the voltage, current, and resistance in a series-parallel circuit.

Calculate the following.

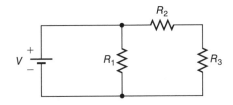

R_1 = 3,456.000 K ohms
R_2 = 40,000.000 K ohms
R_3 = 58,000.000 K ohms

R_total = K ohms

V = 12.0000 volts

VR1 = _____ volts

VR2 = _____ volts

VR3 = _____ volts

I1 = _____ amps

I2 = _____ amps

I3 = _____ amps

Itotal = _____ amps

Instructor's Response

Name: _____   Date: _____

Job Sheet 10: Testing the Accuracy of Voltmeter, Ohmmeter, and Ammeter Meters

- Upon completion of this job sheet, you should be able to check the accuracy of and calibrate an ohmmeter, an ammeter, and a voltmeter.

Ohmmeter Test

Turn the ohmmeter selector switch to the correct range for one of the known resistors. For example, if the resistor is 1,500 ohms, the $R \times 100$ scale will be correct for most meters because it will cause the needle to read about mid-scale. Your instructor will show you the proper scale for the resistors he gives you to test. Place the alligator clip ends together and adjust the meter to 0 by using the 0 adjust knob.

Using several different resistors of known value, record the resistances:

Meter range setting, $R \times$ _____ Resistor value _____ ohms

Meter reading _____ Ohms difference in reading _____ ohms

Meter range setting, $R \times$ _____ Resistor value _____ ohms

Meter reading _____ Ohms difference in reading _____ ohms

Meter range setting, $R \times$ _____ Resistor value _____ ohms

Meter reading _____ Ohms difference in reading _____ ohms

Meter range setting, $R \times$ _____ Resistor value _____ ohms

Meter reading _____ Ohms difference in reading _____ ohms

Meter range setting, $R \times$ _____ Resistor value _____ ohms

Meter reading _____ Ohms difference in reading _____ ohms

Voltmeter Test

In this test, you will use the best voltmeter you have and compare it to others. Your instructor will make provisions for you to take voltage readings across typical low-voltage and high-voltage components. You must perform these tests in the following manner and follow any additional safety precautions from your instructor. Use only meter leads with insulated alligator clips. You and your instructor should ensure that all power to the unit you are working on is turned off. Fasten alligator clips to terminals indicated by your instructor. Make sure that all range selections have been set properly on your meter.

When your instructor has approved all connections, turn the power on and record the measurement.

1. Take a voltage reading of both a high- and low-voltage source using your best meter. (We will call it a "standard" meter.) Your instructor will show you the appropriate terminals.

Record the voltages here:

Low-voltage reading _____ volts High-voltage reading _____ volts

2. Use your other meters such as the VOM meter and the ammeter's voltmeter feature and compare their readings to the standard meter. Record the following:

STANDARD: Low volts reading _____ volts   High volts reading _____ volts

VOM: Low volts reading _____ volts   High volts reading _____ volts

Ammeter: Low volts reading _____ volts   High volts reading _____ volts

Record any differences in the readings here.

VOM reads: _____ volts high, or _____ volts low

Ammeter reads: _____ volts high, or _____ volts low

Ammeter Test

(Follow all procedures indicated in the voltmeter test.)

1. With the power off, connect the electric resistance heater to the power supply.

2. Fasten the leads of the most accurate voltmeter to the power supply directly at the heater.

3. Take an ohm reading with the power off. _____ Ω

4. Clamp the ammeter around one of the conductors leading to the heater.

5. Change the selector switch to read amps. Have your instructor check the connections, and then turn the power on long enough to record the following. (The reason for not leaving the power on for a long time is that the resistance of the heater will change as it heats, and this will change the ampere reading.)

Volt reading _____ volts Ampere reading _____ amps

Volt reading _____ volts ÷ Ohm reading _____ Ω = Ampere reading _____ amps

6. Use a bench meter, if possible, to compare to the clamp-on ammeter.

Bench meter reading _____ amps   Clamp-on ammeter reading _____ amps
Difference in the two _____ amps

Instructor's Response

Chapter 6 Electrical Facility Maintenance

OBJECTIVES

Upon completion of this chapter, you will be able to:

Knowledge-Based

- ☢ Understand and apply OSHA regulations that cover electrical installations.
- ☢ Describe the difference between AC and DC.
- ☢ Correctly identify single-phase and three-phase electrical systems.
- ☢ Correctly identify and select the boxes most commonly used in electrical installations.
- ☢ Correctly identify and select different types of electrical devices and fixtures.
- ☢ Describe the different types of emergency backup systems.

Skill-Based

- ☢ Follow systematic, diagnostic, and troubleshooting partices.
- ☢ Perform tests on smoke alarms, fire alarms, medical alert systems, and emergency exit lighting.
- ☢ Perform tests on GFCI receptacles.
- ☢ Repair and/or replace common electrical devices such as receptacles and switches.
- ☢ Repair and/or replace lighting fixtures and/or bulbs, and ballasts.

Keywords

Voltage   Resistance

Current (or amperage)   Power

Introduction

Typically a facility maintenance technician is responsible for making minor repairs to electrical systems. New installation and additions are typically done by a master licensed electrician. However, when facility maintenance technicians do work with or around electricity, they should follow the safety guidelines outlined below:

1. *Follow the* National Electrical Code *as a standard when making electrical connections and calculating wire sizes and circuit protection.*

2. *Make sure the electrical power supply is shut off at the distribution or entrance panel and locked out or marked in an approved manner.*

3. *Always make sure that the electrical power supply is off on the unit that is being serviced unless electrical energy is required for the service procedure.*

4. *Always keep your body out of contact with damp or wet surfaces when working on live electrical circuits. If you must work in damp or wet areas, make certain that some method is used to isolate your body from these areas.*

5. *Be cautious when working around live electrical circuits. Do not allow yourself to become part of the electrical circuit.*

6. *Use only properly grounded power tools connected to properly grounded circuits.*

7. *Do not wear rings, watches, or other jewelry when working in proximity to live electric circuits.*

8. *Wear shoes with an insulating sole and heel.*

9. *Do not use metal ladders when working near live electrical circuits.*

10. *Examine all extension cords and power tools for damage before using.*

11. *Replace or close all covers on receptacles that house electrical wiring and controls.*

12. *Make sure that the meter and the test leads being used are in good condition.*

13. *Discharge all capacitors with a 20,000-ohm, 4-watt resistor before touching the terminals.*

14. *When attempting to help someone who is being electrocuted, do not become part of the circuit. Always turn the electrical power off or use a nonconductive material to push the person away from the source.*

15. *Keep tools in good condition, and frequently check the insulated handles on tools that are used near electrical circuits.*

Electric Motors

In an electric motor, electric energy is changed to mechanical energy by magnetism, which causes the motor to turn. The motor is based on the principle that like poles of a magnet cause the motor to rotate. This is caused by the repelling action caused by like magnets. To make an electric motor rotation continuous, the magnet field must rotate. This produces a reversal of the poles, or the polarity.

Types of Electric Motors

The industry uses all kinds of AC motors to rotate the many different devices that require rotation in a complete system. Different motors are needed for different tasks because not all motors have the same running and starting characteristics.

Motor Strength

Motor strength is generally used to classify motors into different types. Motors are selected mainly because of the starting torque required for the motor to perform its function.

Motor Speed

$S = (F \times 120)/NP$

S = Speed

F = Flow reversals/second

NP = Number of Poles

Transformers

Transformers are electrical devices that produce an electrical current in a second circuit through electromagnetic induction. Transformers have a primary winding, a core usually made of thin plates of steel laminated together, and a secondary winding. There are step-up and step-down transformers. A step-down transformer contains more turns of wire in the primary winding than in the secondary winding. The voltage at the secondary is directly proportional to the number of turns of wire in the secondary as compared to the number of turns in the primary windings.

Sizing Transformers

Transformers are like many other electrical components. They are not 100 percent efficient. There is a loss between the primary and secondary windings. This loss is considered when sizing transformers for a certain job. Transformers are rated by their primary voltage, secondary voltage, and voltamperes (VAs). System equipment must be considered in transformer sizing along with the transformer rating.

Chapter Review Questions and Exercises

TRUE/FALSE

1. _____ When attempting to help someone who is being electrocuted, you should pull the person away from the source before doing anything else.

2. _____ It is a good idea to always examine all extension cords and power tools before using them.

3. _____ Do not wear rings, watches, or other jewelry when working in proximity to live electric circuits.

4. _____ If a three-prong receptacle is not available then it is on to cut the ground plug off?

5. _____ Always make sure that the electrical power supply is off on the unit that is being serviced unless electrical energy is required for the service procedure.

6. _____ Motor speed is generally used to classify motors into different types.

7. _____ Motors are selected mainly because of the starting torque required for the motor to perform its function.

8. _____ Motors are commonly either open or enclosed.

9. _____ Resistance-start-induction-run motor has only a starting winding and running.

10. _____ An alternating current motor's speed is determined by the frequency of the power supply.

11. _____ The transformer for a residential unit is used to convert line voltage to 24 volts.
12. _____ Some commercial and industrial high-voltage types of equipment use transformers that drop the line voltage from 240 or 120 volts.
13. _____ The transformer is a heating or cooling system that provides the low-voltage power source for the control circuit.
14. _____ Transformers are 100 percent efficient.
15. _____ Transformers are rated by their primary voltage, secondary voltage, and voltamperes (VAs).

COMPLETION

16. _____ motor produces a high-starting torque, which is needed for many applications in the industry.
17. The_____ motor produces a high-starting torque and increases the running efficiency.
18. _____ are stationary inductive devices that transfer electric energy from one circuit to another by induction.

SHORT ANSWER

19. How do electric motors work?

_____

_____

20. Identify each variable in the speed formula listed below.

$S = (F \times 120)/NP$

_____

_____

21. What are shaded pole motors used for?

_____

_____

22. What are the two windings associated with a transformer?

_____

_____

23. What is the purpose of the primary windings of a transformer?

_____

_____

Name: _____   Date: _____

Job Sheet 1: Checking a Smoke Detector

- One of the most common jobs of a facility maintenance technician is to keep smoke detectors in good working order.

Step	Pass	Fail
Press the battery-test button on the unit to make sure the battery is properly connected.		
If the unit has a battery that's more than a year old, replace the battery (see Job Sheet 2).		
Light a candle and hold it approximately 6 inches below the detector so that heated air will rise into the unit.		
If the alarm doesn't sound within 20 seconds, blow out the candle and let the smoke rise into the unit.		
If the alarm still doesn't sound, open the unit up and make sure it is clean and that all electrical connections are solid.		

Note: If, again, the alarm doesn't sound, replace the smoke detector.

Instructor's Response

Name: _____   Date: _____

Job Sheet 2: Replacing the Battery on a Smoke Detector

- When replacing a smoke detector battery, use the following checklist.

Step	Complete
Remove the smoke detector cover, typically by carefully pulling down on the case's perimeter or by twisting the case counterclockwise.	
Locate and remove the battery.	
Replace it with a new one.	
Close the case and test the smoke detector (see Job Sheet 1).	
Read the owner's manual for additional troubleshooting tips and possible adjustments.	

Instructor's Response

Name: _____   Date: _____

Job Sheet 3: Replacing a Receptacle

- Use the following checklist to replace a receptacle.

Step	Complete
Shut off the power at the circuit breaker box.	
Remove the cover plate and test the terminals with the circuit tester. If the tester doesn't light up, there is no power going to the switch.	
Remove the screws that hold the switch in place and pull the switch from the wall.	
Loosen the screws holding the wires to the switch, remove the wires, and remove the switch. In some newer switches, the wires may go directly into the switch, where they are held in place by clamps inside the switch. These switches usually have a slot into which you can insert a small screwdriver to loosen the clamps.	

Instructor's Response

Name: _____   Date: _____

Job Sheet 4: Basic Understanding of Electric Motors

- Upon completion of this job sheet, you should be able to test your basic understanding of electric motors.

MULTIPLE CHOICE

1. When an alternating current is applied to the running windings of a split-phase electric motor, an alternating current:
   A. with the same polarity is induced in the rotor.
   B. with reverse polarity is induced in the rotor.
   C. with the same polarity is induced in the start winding.
   D. circuit to the start winding is opened by the electronic relay.

2. A two-pole split-phase motor operates at a speed just under:
   A. 1,800 rpm.
   B. 3,600 rpm.
   C. 4,800 rpm.
   D. 5,200 rpm.

3. Start windings are placed _____ the run windings.
   A. outside
   B. inside
   C. between
   D. in series with

4. The start windings have:
   A. fewer turns than the run windings.
   B. more turns than the run windings.
   C. the same number of turns as the run windings.

5. A four-pole split-phase motor runs _____ a two-pole split-phase motor.
   A. faster than
   B. at the same speed as
   C. slower than

6. When a typical split-phase motor reaches approximately 75 percent of its operating speed, the start winding circuit is opened by a:
   A. circuit breaker.
   B. capacitor.
   C. centrifugal switch.
   D. dual-voltage switch.

7. The electronic relay is used with some electric motors to:
   A. provide current to the start windings.
   B. relay current to the start capacitor.
   C. provide current to the run capacitor.
   D. open the circuit to the start windings after the motor has started.

8. The amount by which the current leads or lags the voltage in an AC circuit is known as the:
   A. phase angle.
   B. electronic relay.
   C. inductive angle.
   D. starting torque.

9. The shaded-pole motor:
   A. has little starting torque.
   B. has excessive starting torque.
   C. has greater efficiency than most split-phase motors.
   D. is designed for three-phase operation.

Instructor's Response

Name: _____   Date: _____

Job Sheet 5: Basic Understanding of Transformers

- Upon completion of this job sheet, you should be able to test your basic understanding of transformers. Circle the letter that indicates the correct answer.

MULTIPLE CHOICE

1. The power-handling capabilities of transformers are measured in units of:
   A. volts.
   B. amps.
   C. voltamperes.
   D. watts.

2. The turns ratio of a transformer is defined as:
   A. the number of primary turns multiplied by the secondary turns.
   B. the number of primary turns divided by the secondary turns.
   C. the number of secondary turns multiplied by the primary turns.
   D. the number of secondary turns divided by the primary turns.

3. A transformer in which the secondary voltage is less than the primary voltage is:
   A. a step-down transformer.
   B. a step-up transformer.
   C. a center-tap transformer.
   D. a power transformer.

4. A basic transformer in its normal operation can:
   A. step up voltage.
   B. step down voltage.
   C. step up current.
   D. all of these.

5. A step-up transformer in its normal operation:
   A. steps up the secondary voltage.
   B. steps down the secondary current.
   C. maintains the output power close to the input power.
   D. all of these.

6. The formula for the current ratio in a transformer is:
   A. opposite to the turns ratio.
   B. the same as the turns ratio.
   C. less than the turns ratio.
   D. more than the turns ratio.

7. If a transformer has 120 VAC applied to the primary with 200 turns and a secondary with 100 turns, the secondary voltage is:

   A. 60 VAC.

   B. 120 VAC.

   C. 240 VAC.

   D. none of these.

8. If a transformer is rated 120 VAC, 500 milliamperes maximum, its power handling capacity is:

   A. 5 amps AC.

   B. 120 volts AC.

   C. 60 voltamperes.

   D. 600 voltamperes.

9. Power transformers outside your home are:

   A. the step-down type.

   B. the step-up type.

   C. the impedance matching type.

   D. the phase shifting type.

10. Power transformers outside your home are designed to:

   A. save volts.

   B. save amps.

   C. save resistance.

   D. save power.

Instructor's Response

Name: _____   Date: _____

Job Sheet 6: Basic Understanding of Transformers

- Upon completion of this job sheet, you should be able to test your basic understanding of transformers.

TRUE/FALSE

1. _____ Transformers can step up or step down AC or DC voltages.
2. _____ In a transformer circuit, the load is connected to the primary.
3. _____ Regardless of whether the transformer is step-up or step-down, power in the primary equals the power in the secondary if there is 100 percent efficiency.

MULTIPLE CHOICE

4. Increasing the number of turns of wire on the secondary of a transformer will:
   A. decrease the output voltage.
   B. increase the output voltage.
   C. have no effect on the output voltage.
   D. increase the primary input voltage.

5. Increasing the number of turns of wire on the secondary of a transformer will:
   A. increase the secondary current.
   B. decrease the secondary current.
   C. have no effect on the secondary current.
   D. increase the primary current.

6. A transformer has 430 turns in the primary and 200 turns in the secondary. What is the primary-to-secondary turns ratio?
   A. 230:1
   B. 2.15:1
   C. 1:2.15
   D. 1:30

COMPLETION

7. The ability to transfer energy from one coil to another is called _____.

8. Transformer efficiency percentage is the product of _____ power divided by _____ power times 100.

9. A transformer coil connected to a source of AC voltage is the _____ winding.

Instructor's Response

Chapter 7   Carpentry

Upon completion of this chapter, you will be able to:

Knowledge-Based

- ⊗ Describe the general properties of hardwood and softwood commonly used by facilities maintenance technicians.
- ⊗ Describe the effects of moisture content on different wood products.
- ⊗ Perform estimating and takeoff quantities for simple one-step carpentry projects.
- ⊗ Correctly identify and select engineered products, panels, and sheet goods.
- ⊗ Correctly identify framing components.

Skill-Based

- ⊗ Perform interior carpentry maintenance.
- ⊗ Perform exterior carpentry maintenance.

Keywords

Hardwood	Plies	Multispur bits
Softwood	Softboard	Faceplate markers
Green lumber	Back miter	Striker plate
Engineered panels	Boring jig	

Introduction

All facility maintenance technicians have to deal with structural repairs and construction. This could be in the form of repairing a wall that has been damaged or adding a new wall to divide an existing space. Either case will require the technician to estimate the amount of time and determine the materials and tools needed to accompany the task.

Estimating the Amount of Aluminum and Vinyl Siding

Aluminum and vinyl siding panels are typically sold by the square. To determine the amount of siding needed for a project, first determine the amount of wall area to be covered. Second, add 10 percent of the wall area calculated for waste. Third, divide the amount calculated in the previous step by 100. For example, to determine the amount of siding needed for the doctor offices shown in Figure 7-1, first identify that the exterior wall height for the structure is 9'6".

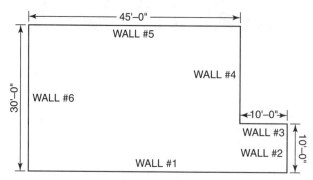

FIGURE 7-1: Estimating siding.

1. Determine the wall area to be covered.

Wall #1
Length = length of wall #5 + length wall #3
Length = 45 ft. + 10 ft.
Length = 55 ft.

Because the height of the wall is 9 ft. 6 in., or 9.5 ft., the area is calculated as

Area1 = 55 ft. · 9.5 ft.
Area1 = 522.5 sq. ft.

Wall #2
Area2 = 10 ft. · 9.5 ft.
Area2 = 95 sq. ft.

Wall #3
Area3 = 10 ft. · 9.5 ft.
Area3 = 95 sq. ft.

Wall #4
Area4 = 20 ft. · 9.5 ft.
Area4 = 190 sq. ft.

Wall #5
Area5 = 45 ft. · 9.5 ft.
Area5 = 427.5 sq. ft.

Wall #6
Area6 = 30 ft. · 9.5 ft.
Area6 = 285 sq. ft.

Total Square Footage
Area Total = Area1 + Area2 + Area3 + Area4 + Area5 + Area6
Area Total = 522.5 sq. ft. + 95 sq. ft. + 95 sq. ft. + 190 sq. ft. + 427.5 sq. ft. + 285 sq. ft.
Area Total = 1,615 sq. ft.

2. Determine the amount of waste.

Allowing for waste

Waste Allowance = Area Total · 0.10

Waste Allowance = 161.5 sq. ft.

3. Determine the number of squares.

Total Wall Area with Waste = 1,615 sq. ft. + 161.5 sq. ft.

Total Wall Area with Waste = 1,776.5 sq. ft.

Number of Squares = Total Wall Area with Waste / 100

Number of Squares = 1,778.5 / 100

Number of Squares = 17.765 or 18

Estimating Suspended Acoustical Ceilings

The ceiling of choice for most commercial construction projects is a suspended acoustical ceiling. The amount of ceiling tiles needed for this type of ceiling can be determined by simply multiplying the length of the room to be covered by the width of the room and then dividing that by the area of the ceiling tile to be used. For example, calculate the number of ceiling tiles needed for a room 10 × 15 feet, using 2 × 4 foot ceiling tiles.

1. Calculate the total area of the ceiling to be covered.

Ceiling Area = Length of Room × Width of Room

Ceiling Area = 10 ft. × 15 ft.

Ceiling Area = 150 sq. ft.

2. Calculate the total area of the ceiling tile.

Ceiling Tile Area = Length of Ceiling Tile × Width of Ceiling Tile

Ceiling Tile Area = 2 ft. × 4 ft.

Ceiling Tile Area = 8 sq. ft.

3. Determine the number of ceiling tiles to use.

Number of Ceiling Tiles = Ceiling Area / Ceiling Tile Area

Number of Ceiling Tiles = 150 sq. ft. / 8 sq. ft.

Number of Ceiling Tiles = 18.75 or 19

Chapter Review Questions and Exercises

TRUE/FALSE

1. _____ When calculating the amount of siding needed, a waste factor of 15 percent is always used.

2. _____ Softwood is not used in construction and is typically used in the production of paper.

SHORT ANSWER

3. How does hardwood differ from softwood?

_____

_____

4. List two applications for hardwood.

_____

_____

5. Calculate the number of ceiling tiles that will be needed for a room 15 × 15 feet, using 2 × 4 foot ceiling tiles.

_____

_____

Name: _____   Date: _____

Job Sheet 1: Estimating Drywall

Before starting a remodeling, new construction, or repair project, it is necessary to determine the amount of materials that will be needed to complete the assigned task. In most cases, this can be done at the home builder's center by simply providing the necessary dimensions. However, material estimation can be easily accomplished with a little practice. In many cases, there are free estimating forms and programs on the Internet that can be used to accomplish this. Below is a sample estimating form in which the amount of drywall can be estimated.

Areas	Height	Width	Height × Width	Total Area
Wall				
Ceiling				
Excluded Areas				
		Base	**(Base /2) · Height**	
Sloping Wall				

Sheet Rock Size	4 × 8	4 × 9	4 × 10	4 × 12
Area	32	36	40	48

Total Dry Wall Area   = _____
([Wall + Ceiling] − Excluded Areas + Sloping Walls)

Total Drywall Required   = _____
([Total Dry Wall Area]/Sheet Rock Area)

Instructor's Response

Name: _____   Date: _____

Job Sheet 2: Using the Drywall Estimating Form

Use the drywall estimating form provided below to determine the amount of drywall needed to cover a 10 × 12 foot wall and a 10 × 15 foot ceiling.

Areas	Height	Width	Height × Width	Total Area
Wall				
Ceiling				
Excluded Areas				
		Base	(Base /2) · Height	
Sloping Wall				

Sheet Rock Size	4 × 8	4 × 9	4 × 10	4 × 12
Area	32	36	40	48

Total Dry Wall Area   = _____
([Wall + Ceiling] − Excluded Areas + Sloping Walls)

Total Drywall Required   = _____
([Total Dry Wall Area]/Sheet Rock Area)

Instructor's Response

Name: _____   Date: _____

Job Sheet 3: Estimating Ceiling Tiles

For the sketch provided below, determine the number of ceiling tiles needed.

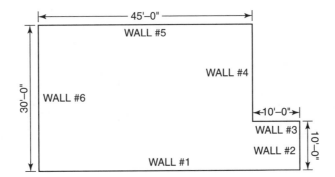

Instructor's Response

Name: _____   Date: _____

Job Sheet 4: Determining the Joist Length of a Cantilevered Joist

Determining the Joist Length of a Cantilevered Joist

Upon completion of this job sheet you will be able to determine the length of a cantilevered joist.

Calculation

Determine the length of the joist for the cantilevered joist shown in the illustration below.

Framing plan

Extended over foundation

Overall dimension

Over All Dimension	Extended over Foundation
	4 foot
	6 foot
	3 foot
	5 foot–6 inches
	10 foot

Instructor's Response

Name: _____   Date: _____

Job Sheet 5: Identifying Components of Kitchen Cabinets

Using the Correct Nail

Upon completion of this job sheet you will be able to identify various components of kitchen cabinets.
Identify the following.

Kitchen Cabinets

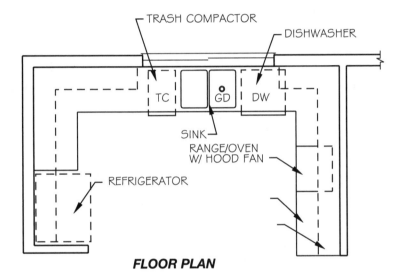

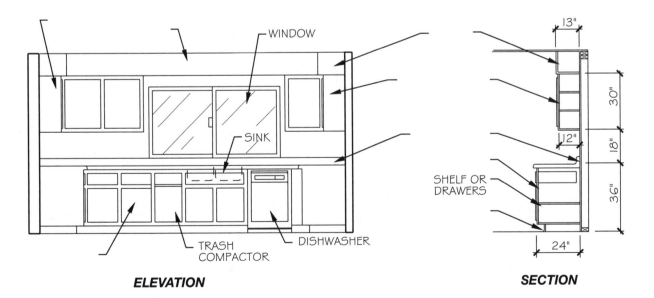

Instructor's Response

Name: _____   Date: _____

Job Sheet 6: Types of Solid and Built-Up Headers

Solid and Built-Up Headers

Upon completion of this job sheet you will be able to identify the various types of solid and built-up header typically used in framing.

Identify the types of header pictured below.

Identification

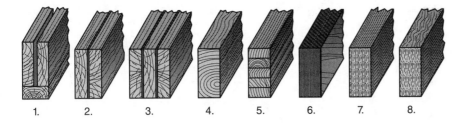

1.   2.   3.   4.   5.   6.   7.   8.

Instructor's Response

Name: _____   Date: _____

Job Sheet 7: Window Framing

Window Framing

Upon completion of this job sheet you will be able to identify the various components of a window opening. Identify the components pictured below.

Identification

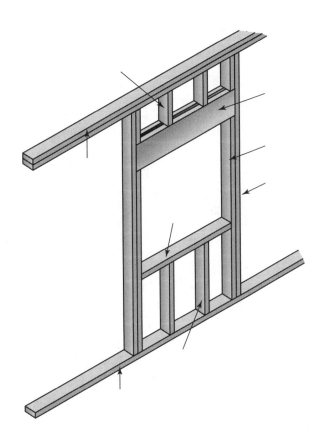

Instructor's Response

Name: _____   Date: _____

Job Sheet 8: Door Framing

Door Framing

Upon completion of this job sheet you will be able to identify the various components of a door opening. Identify the components pictured below.

Identification

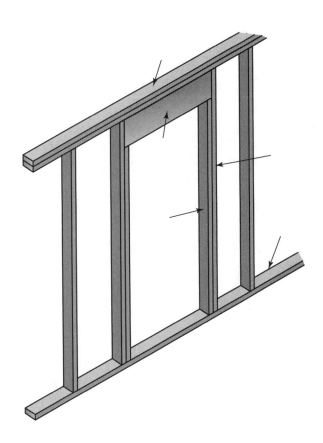

Instructor's Response

Name: _____   Date: _____

Job Sheet 9: Identifying Truss Components

Truss Component Identification

Upon completion of this job sheet you will be able to identify the various trusses used in a residential construction.

For each of the trusses listed create a quick sketch.

Identification

Scissors truss

Modified fink

Camel back pratt

Saw-tooth truss

Instructor's Response

Name: _____   Date: _____

Job Sheet 10: Truss Component Identification

Truss Component Identification

Upon completion of this job sheet you will be able to identify the various components of a roof truss. Identify the various components of the truss pictured below.

Identification

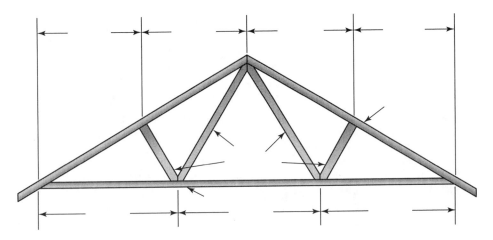

Instructor's Response

Name: _____   Date: _____

Job Sheet 11: Hanging an Interior Door

Hanging Interior Doors

Upon completion of this job sheet you will be able to identify the steps for hanging an interior door.

In the space provided below list the steps for hanging an interior door.

Procedure

Instructor's Response

Name: _____ Date: _____

Job Sheet 12: Installing Windows Job

Installing Windows

Upon completion of this job sheet you will be able to identify the steps for installing windows.

In the space provided below list the steps for installing windows.

Procedure

Instructor's Response

Name: _____   Date: _____

Job Sheet 13: Estimating Shingles

Estimating Shingles

Upon completion of this job sheet you will be able to estimate the amount of shingles needed to roof a house.

Estimate the amount of shingles needed to roof a house, if the fascia boards measure 55′ and the length of the rafter 25′.

Estimate

Instructor's Response

Name: _____   Date: _____

Job Sheet 14: Installing Asphalt Shingles

Installing Asphalt Shingles

Upon completion of this job sheet you will be able to describe the method used to install asphalt shingles.

In the space provided below list the steps for installing asphalt shingles.

Procedure

Instructor's Response

Name: _____ Date: _____

Job Sheet 15: Vinyl Siding

Repairing Siding

The following steps outline the procedure for installing horizontal siding:

Step #1: First determine the sidings exposure around any windows. The siding should have equal exposure both above and below the window's sill.

Step #2: Install a starter strip of the same thickness and width of the siding at the headlap fastened along the bottom edge of the sheathing.

Step #3: From this first chalk line, lay out the desired exposures on each corner board and each side of all openings.

Step #4: Install the siding as per the manufacturer's recommendations, staggering the butt joints in adjacent courses as far apart as possible.

Step #5: When applying a course of siding, start from one end and work toward the other end.

Step #6: Siding is fastened to each bearing stud or about every 16 inches.

The following steps outline the procedure for installing vertical tongue-and-groove:

Step #1: Slightly back-bevel the ripped edge.

Step #2: Fasten a temporary piece on the other end of the wall projecting below the sheathing by the same amount.

Step #3: Apply succeeding pieces by toe-nailing into the tongue edge of each piece.

Step #4: To cut the piece to fit around an opening, first fit and tack a siding strip in place where the last full strip will be located.

Step #5: Next, use a scrap block of the siding material, about 6 inches long, with the tongue removed.

Step #6: Continue the siding by applying the short lengths across the top and bottom of the opening as needed.

Step #7: Fit the next full-length siding piece to complete the siding around the opening.

Step #8: Remove the piece and the scrap blocks from the wall.

The following steps outline the procedure for installing panel siding:

Step #1: Install the first piece with the vertical edge plumb.

Step #2: Apply the remaining sheets in the first course in like manner.

The following steps outline the procedure for installing wood shingles and shakes:

Step #1: Fasten a shingle on both ends of the wall with its butt about 1 inch below the top of the foundation.

Step #2: Fill in the remaining shingles to complete the undercourse.

Step #3: Apply another course on top of the first course.

Step #4: To apply the second course, snap a chalk line across the wall at the shingle butt line.

The following steps outline the procedure for installing horizontal vinyl siding:

Step #1: Snap a level line to the height of the starter strip all around the bottom of the building.

Step #2: Cut the corner posts so they extend ¼ inch below the starting strip.

Step #3: Cut each j-channel piece to extend, on both ends, beyond the casing and sills a distance equal to the width of the channel face.

Step #4: On both ends of the top and bottom channels, make ¾ inch cuts at the bends leaving the tab attached.

Step #5: Snap the bottom of the first panel into the starter strip.

Step #6: Install successive courses by interlocking them with the course below and staggering the joints between courses.

Step #7: To fit around a window, mark the width of the outout, allowing ¼ inch clearance on each side.

Step #8: Panels are cut and fit over windows in the same manner as under them.

Step #9: Install the last course of the siding panel under the soffit in a manner similar to fitting under a window.

The following steps outline the procedure for installing vertical vinyl siding:

Step #1: Measure and lay out the width of the wall section for siding pieces.

Step #2: Cut the edge of the first panel nearest the corner.

Step #3: Install the remaining full strips making sure there is ¼ inch gap at the bottom.

Installing Vinyl Siding

Upon completion of this job sheet you will be able to describe the method used to install vinyl siding.

In the space provided below list the steps for installing vinyl siding.

Procedure

Instructor's Response

Name: _____   Date: _____

Job Sheet 16: Installing Wood Shingles and Shakes

Installing Wood Shingles and Shakes

Upon completion of this job sheet you will be able to describe the method used to install wood shingles and shakes.

In the space provided below list the steps for installing wood shingles and shakes.

Procedure

Instructor's Response

Name: _____   Date: _____

Job Sheet 17: Determining the Rise and Run of a Set of Stairs

Determining the Rise and Run of a Set of Stairs

Upon completion of this job sheet you will be able to determine the rise and run of a set of stairs.

To determine the rise and run of a set of stairs the following steps should be used.
- Total rise = Floor to ceiling height + floor joist + depth of floor covering
- Number or risers = Divide total rise by maximum rise ($7^3/_4$″)
- Number of treads = Number of risers − 1
- Total run = Tread width × number of treads

Stairs

Determine the rise and run for a set of stairs connecting two floors if the distance between the floors is 9 feet. The second floor is framed using 2 × 10′s and has a 23/32 sub floor with a 5/8″ wood floor over that.

Instructor's Response

Name: _____   Date: _____

Job Sheet 18: Stair Terminology

Stair Terminology

Upon completion of this job sheet you will be able to describe some of the parts of a set of stairs.

Define the following:

Stairs

Stringer or stair jack

Kick block or kicker

Headroom

Handrail

Guardrail

Instructor's Response

Chapter 8 Surface Treatments

Upon completion of this chapter, you will be able to:

Knowledge-Based

✪ Identify and select proper surface finishes.

✪ Identify and select proper finishing tools for different types of finishes.

Skill-Based

✪ Prepare surface and site properly for finishing, including sanding, caulking, and covering exposed surfaces.

✪ Apply paint using roller and brush according to manufacturer and job specifications.

✪ Apply paint using a paint sprayer according to manufacturer and job specifications.

✪ Clean and store paint materials including brushes, rollers, thinners, and spray guns according to manufacturer's specifications and OSHA regulations.

Keywords

Mask	Kill spot
Mural	Booking

Introduction

The first thing that anyone notices when he or she enters into a facility is the wall finish. Therefore it is essential that the facility maintenance technicians have a good grasp on surface treatments, the different types, and the maintenance required to keep them in pristine condition.

Before attempting to repair and/or install a surface treatment it is extremely important that the facility maintenance technician be able to estimate the amount and type of materials needed to complete the job. The correct estimation not only saves time but it also allows the technician to do a better job. This is due in part to limiting the number of times in which the technician has to start and stop.

Chapter Review Questions and Exercises

TRUE/FALSE

1. _____ The proper application of paint can change the look of a room at very little cost.

2. _____ Gloss paints are oil based and include resin to give them a hardwearing quality. The higher the gloss level, the lower the shine and easier it will be to maintain.

3. _____ Pressure washers will blast away many years worth of dirt and grime in a short time with minimal effort.

4. _____ Paint brushes come in a variety of shapes, such as angular, flat, and oval. Sizes range from 1 inch to 4 inches in width.

SHORT ANSWER

5. List the two categories of paint brushes.

_____

_____

6. How do you estimate the amount of wallpaper necessary to complete a room?

_____

_____

7. What are the basic tools needed to install wallpaper?

_____

_____

8. How does the wall preparation necessary for wallpaper differ than the wall preparation necessary for paint?

_____

_____

9. Why is it necessary to purchase more wallpaper than required when doing a wallpaper job?

_____

_____

10. What is the procedure for estimating wallpaper?

_____

_____

Name: _____   Date: _____

Job Sheet 1: Estimating the Amount of Paint Needed

- The form provided below is for estimation purposes only. Before purchasing the paint, always consult a painting professional at your paint dealer to double-check your answer.

Width × Height = Area		
1.	Width of Walls (added together)	Room Height	Total Wall Surface Area

Width × Height = Area		
2.	Window Height	Window Width	Window Surface Area
3.	Repeat Step 2 for Each Window	Total Window Surface Area	

Width × Height = Area		
4.	Door Height	Door Width	Door Surface Area
5.	Repeat Step 4 for Each Door	Total Door Surface Area	

Step 3 + Step 5		
6.	Total from Step 3	Total from Step 5	Total Area Not to Paint

Step 1 − Step 6		
7.	Total Step 1	Total Step 6	Total Area to Paint

Step 7 ÷ Spread Rate		
8.	Total Step 7	Spread Rate	Amount of Paint/Coat

Instructor's Response

Name: _____   Date: _____

Job Sheet 2: Estimating the Amount of Paint Practice

- Use the drawing provided below to estimate the amount of paint necessary to paint the room if the wall height is 8 feet.

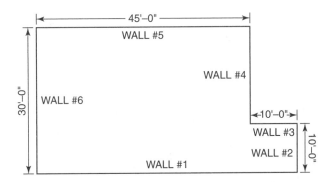

Width × Height = Area		
1.	Width of Walls (added together)	Room Height	Total Wall Surface Area
Width × Height = Area		
2.	Window Height	Window Width	Window Surface Area
3.	Repeat Step 2 for Each Window	Total Window Surface Area	
Width × Height = Area		
4.	Door Height	Door Width	Door Surface Area
5.	Repeat Step 4 for Each Door	Total Door Surface Area	
Step 3 + Step 5		
6.	Total from Step 3	Total from Step 5	Total Area Not to Paint
Step 1 − Step 6		
7.	Total Step 1	Total Step 6	Total Area to Paint
Step 7 ÷ Spread Rate		
8.	Total Step 7	Spread Rate	Amount of Paint/Coat

Instructor's Response

Name: _____   Date: _____

Job Sheet 3: Estimating the Amount of Wallpaper Practice

- Use the drawing provided below to estimate the amount of wallpaper necessary to wallpaper the room if the wall height is 8 feet.

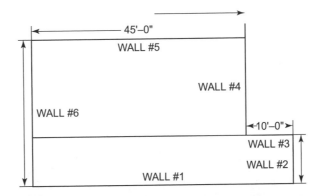

Width × Height = Area		
1. Width of Walls (added together)	Room Height	Total Wall Surface Area
Width × Height = Area		
2. Window Height	Window Width	Window Surface Area
3. Repeat Step 2 for Each Window	Total Window Surface Area	
Width × Height = Area		
4. Door Height	Door Width	Door Surface Area
5. Repeat Step 4 for Each Door	Total Door Surface Area	
Step 3 + Step 5		
6. Total from Step 3	Total from Step 5	Total Area Not to Wallpaper
Step 1 − Step 6		
7. Total Step 1	Total Step 6	Total Area to Wallpaper

Instructor's Response

Chapter 9   Plumbing

OBJECTIVES

Upon completion of this chapter, you will be able to:

Knowledge-Based

- ❂ Identify and select basic plumbing tools for a specific application.
- ❂ Identify and select the proper type of plastic piping.
- ❂ Identify and select the proper type of copper tubing.
- ❂ Identify and select the proper type of metallic pipe.
- ❂ Identify and select the proper type of pipe fitting.
- ❂ Identify and select pipe hangers and supports.
- ❂ Identify, select, and apply caulking.
- ❂ Identify and apply caulking.

Skill-Based

- ❂ Correctly measure and cut copper tubing.
- ❂ Fabricate plastic pipe with correct fittings to correct dimensions as required for job without any leaks.
- ❂ Assemble compression fitting without any leaks.
- ❂ Clean and replace traps, drains, and vents including use of sink snake or rod to clean drain lines.
- ❂ Caulk and seal fixtures according to manufacture's specifications.
- ❂ Fabricate and solder copper pipe with correct fitting as required for job without any leaks.
- ❂ Test and set hot water temperature according to manufacturer's specifications.
- ❂ Follow and apply all national and local building codes.
- ❂ Locate and repair leaks in pipes and fixtures.
- ❂ Install shower seals.
- ❂ Repair, replace, and/or rebuild plumbing fixtures and connections to job specifications without any leaks.

Keywords

Thermostat

Introduction

Because plumbing is an essential part of any facility for providing water and removing waste products, it is critical that the facility maintenance technician have a basic understanding of fundamental plumbing techniques and principles. This understanding allows the technician to be more effective in unclogging drains and/or providing minor plumbing fixes. Having a good understanding of the tools used in to repair plumbing systems is essential to correctly and safely completing a particular repair.

Chapter Review Questions and Exercises

TRUE/FALSE

1. _____ The right tool can help a facility maintenance technician complete a task more efficiently and safely.

2. _____ The plumbing trade uses only specialty tools.

SHORT ANSWER

3. Why is it important to use the correct tool for a specific task?

_____

_____

4. What are swaging tools used for?

_____

_____

5. What are copper bending tools used for?

_____

_____

6. A(n) _____ is an enlarged end of a pipe or fitting that receives a pipe end or fitting and that may also be called a bell or a hub.

7. A pipe that receives discharge from a fixture(s) is a _____.

8. A(n) _____ is a fitting used to create an offset; it is also called a bend.

9. A(n) _____ is a fitting having internal threads and screws over a male fitting.

10. A type of offset fitting that has one end with the same outside diameter as a connecting pipe or fitting is called a _____.

11. A(n) _____ is an offset made in tubing on a job site and a manufactured offset fitting.

12. An enlarged end of a pipe or fitting that receives a pipe end or fitting and that may also be called a hub or socket is a _____.

13. A(n) _____ is a pipe dedicated to provide airflow so a drainage system can breathe.

14. A(n) _____ is a type of vacuum breaker that is commonly used on a water heater that is piped with a side inlet connection.

15. Unobstructed vertical space from a device outlet to a point where water could backflow into a piping system is called _____.

16. _____ is cleaning a piping system with air or water pressure.

17. A designated raised portion of a valve and device used in place of a handle to operate with a wrench or tool is a(n) _____.

Name: _____   Date: _____

Job Sheet 1: The Plumber's Toolbox

- Upon completion of this job sheet, you should be able to identify the specialty tools used in all plumbing trades.
- This job sheet is a checklist of the basic tools used by all plumbing trades.

Tool	Present	Condition
Pipe Nipple Extracting Set	☐	_____
Thread Tapping Tool	☐	_____
Inside Plastic Pipe Cutter	☐	_____
Basin Wrench	☐	_____
Mini-Hacksaw	☐	_____
Nail-Puller	☐	_____
Smooth-Jaw Pipe Wrench	☐	_____
Plastic Pipe Saw	☐	_____
Needle-Nosed Pliers	☐	_____
Locking Pliers	☐	_____
6" Combination Pliers	☐	_____
Copper Flaring Tool	☐	_____
Copper Tubing Bender	☐	_____
Flexible Tubing Cutter	☐	_____
Copper Swaging Tool	☐	_____
Basket Strainer Tool or Internal Wrench	☐	_____
Multi-Purpose Knife/Pliers Tool	☐	_____
Carpenter Square	☐	_____
Metal Stud Punch	☐	_____
Strap Pipe Wrench	☐	_____
Chain Pipe Wrench	☐	_____
Miniature Hacksaw	☐	_____
Wallboard Saw	☐	_____
Flexible Tubing Crimping Tool	☐	_____
Ball Peen Hammer	☐	_____
Inside Plastic Pipe Cutter	☐	_____
Cast Iron Chain Cutters	☐	_____
Internal Cast Iron Cutters	☐	_____

Instructor's Response

Name: _____   Date: _____

Job Sheet 2: Ordering a Plumbing Tee

- Upon completion of this job sheet, you should be able to order a plumbing tee.
- For each tee, correctly provide the ordering information.

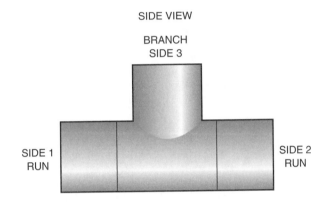

SIDE VIEW

BRANCH
SIDE 3

SIDE 1
RUN

SIDE 2
RUN

Side 1	Side 2	Side 3	Order as
1/2″	1/2″	1/2″	
1/2″	1/2″	3/4″	
3/4″	3/4″	3/4″	
3/4″	3/4″	1/2″	
3/4″	1/2″	1/2″	
3/4″	1/2″	3/4″	
3/4″	3/4″	1″	
1″	1″	1″	
1″	1″	3/4″	
1″	1″	1/2″	
1″	3/4″	3/4″	
1″	3/4″	1/2″	
1″	1/2″	1/2″	

Instructor's Response

Name: _____   Date: _____

Job Sheet 3: Fitting Identification

- Upon completion of this job sheet, you should be able to correctly identify various fittings used in plumbing.
- For each fitting, write the name under its picture.

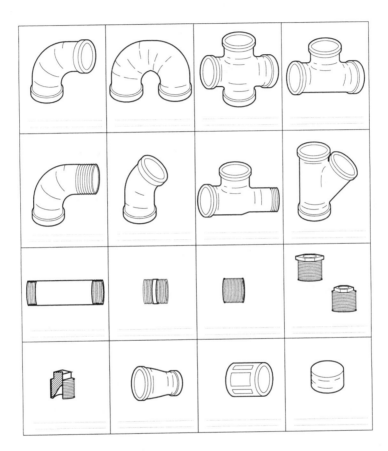

Instructor's Response

Name: _____   Date: _____

Job Sheet 4: Valve Identification

- Upon completion of this job sheet, you should be able to correctly identify various valves used in plumbing.
- For each valve, give a brief description of what that valve is used for.

Type	Residential Uses
Ball Valve	
Gate Valve	
Stop Valve	
Stop and Waste Valve	
Gas Cock	

Instructor's Response

Name: _____   Date: _____

Job Sheet 5: Valve Identification

- Upon completion of this job sheet, you should be able to correctly identify various valves used in plumbing.
- For each valve, give a brief description of what that valve is used for.

Type	Use
Pressure-Reducing Valve	
Check Valve	
Vacuum Breaker	
Vacuum Relief Valve	
Relief Valve	
Reduced Pressure Zone Valve	
Double-Check Valve Assembly	

Instructor's Response

Name: _____   Date: _____

Job Sheet 6: Faucet Identification

- Upon completion of this job sheet, you should be able to identify the various parts of a faucet.
- Identify the various parts of the faucet shown below.

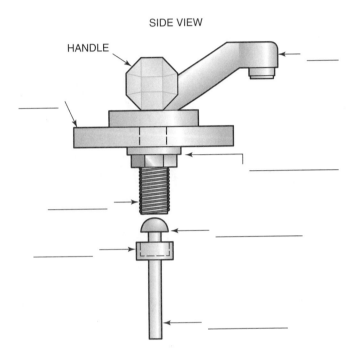

Instructor's Response

Name: _____   Date: _____

Job Sheet 7: Bidet Drain Assembly

- Upon completion of this job sheet, you should be able to identify the various parts of a bidet drain assembly.
- Identify the various parts of the bidet drain assembly shown below.

SIDE VIEW

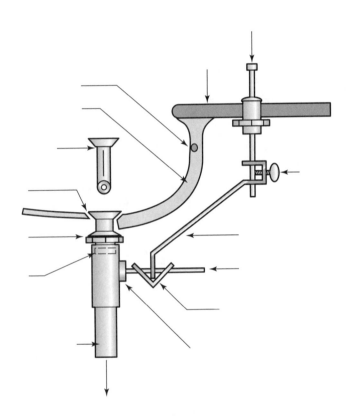

Instructor's Response

Chapter 10 Heating, Ventilation, and Air-Conditioning Systems

OBJECTIVES

Upon completion of this chapter, you will be able to:

Knowledge-Based

- ☺ Explain the importance of properly installing air filters
- ☺ List the three common types of furnaces used

Skill-Based

- ☺ Perform general maintenance procedures including:
  - ☺ General maintenance on a furnace
  - ☺ Tightening and/or replacing belts
  - ☺ Adjusting and/or replacing pulleys
  - ☺ Replacing filters on HVAC units
- ☺ Maintain the heat source on gas-fired furnaces.
- ☺ Perform general maintenance of hot water or steam boilers.
- ☺ Perform general maintenance of an oil burner and boiler.
- ☺ Repair and replace electrical devices, zone valves, and circulator pumps.
- ☺ Light a standing pilot.
- ☺ Perform general maintenance of a chilled water system.
- ☺ Clean coils.
- ☺ Lubricate motors.
- ☺ Follow systematic diagnostic and troubleshooting practices.
- ☺ Maintain and service condensate systems.
- ☺ Replace through-the-wall air conditioners.

Keywords

Carbon monoxide   Carbon dioxide

Introduction

Because of EPA requirements dealing with Freon, all states require that any person handling certain types of Freon must be certified. However, as a facility maintenance technician, regardless of your certification, you will be responsible for some of the basic HVAC maintenance. Typically, this will include cleaning the coil, draining and cleaning condensation lines and pans, filter replacement, and so on.

Chapter Review Questions and Exercises

TRUE/FALSE

1. _____ Damaged and worn belts do not have to be replaced immediately to help ensure the satisfactory, continued operation of the system.

2. _____ Never attempt to adjust the pitch on a variable-pitch pulley to tighten the belt tension.

SHORT ANSWER

3. Why is it important to ensure that pulleys are properly aligned?

_____

_____

4. What are the three types of furnaces used?

_____

_____

5. What is the purpose of HVAC?

_____

_____

Name: _____   Date: _____

Job Sheet 1: HVAC Maintenance Tool Checklist

- Before attempting to perform maintenance on any HVAC equipment always check to make sure that you have all the necessary tools and that they are in good working condition.

Tool	Present	Condition
Tool Box	☐	_____
Flat-Head Screwdriver	☐	_____
Phillips-Head Screwdriver	☐	_____
Common Pliers (6")	☐	_____
Channel Lock Pliers (11")	☐	_____
Lineman Pliers (8")	☐	_____
Needle-Nose Pliers (8")	☐	_____
8" Side Cutting Pliers	☐	_____
Adjustable Wrench (10")	☐	_____
Adjustable Wrench (8")	☐	_____
Ball Peen Hammer (8 oz.)	☐	_____
Drop Light (25')	☐	_____
Tape Measure (10')	☐	_____
Allen Wrench Set $9^{1}/_{2}$"	☐	_____
Box and Open End Combination Wrench Set ($^{3}/_{8}$" to 1")	☐	_____
Safety Glasses	☐	_____
$^{3}/_{8}$" Socket Set Ratchet with Extensions	☐	_____

Instructor's Response

Name: _____   Date: _____

Job Sheet 2: Determining the Correct Belt Size

- Use the chart below to determine the size of the replacement belt for a system in which the center-to-center distance is 20 inches. Pulley #1 has a diameter of 12 inches and Pulley #2 has a diameter of 10 inches.

1. Center to Center Distance

2. Center to Center Distance x 2

3. Pulley Diameter #1   Pulley Diameter 2

4. Pulley Diameter #1 + Pulley Diameter 2

5. Step 3 ÷ 2

6. Step 5 x 3.14

7. Step 6 + Step 2

Instructor's Response

Name: _____   Date: _____

Job Sheet 3: Checking a Package Sequencer

Upon completion of this job sheet, you should be able to check a package sequencer by energizing the operating coil and checking the contacts of the sequencer with an ohmmeter.

PROCEDURES

1. Turn the VOM function switch to AC voltage and the range selector switch to the 50 Volts or higher scale.
2. With the power off, fasten one VOM lead to one of the 24-V power supply leads and the other VOM lead to the other supply lead.
3. Turn the power on and record the voltage _____V. It should be very close to 24 V.
4. Turn the power off.
5. Fasten the two leads from the 24-V power supply to the terminals on the sequencer that energize the contacts. They are probably labeled heater terminals.
6. Turn the ohms selector range switch to R x 1. Fasten the VOM leads on the terminals for one of the electric heater contacts.
7. Turn the 24-V power supply on. You should hear a faint audible "click" each time a set of contacts makes or breaks.
8. After about 3 minutes, check each set of contacts with the VOM to see that all sets are made. You may want to disconnect the power supply and allow all contacts to open and repeat the test to determine the sequence in which the contacts close.
9. Turn off the power and disconnect the sequencer.

QUESTIONS

1. What type of device is inside the sequencer to cause the contacts to close when voltage is applied to the coil?

   _____

   _____

2. What type of coil is inside the sequencer? (magnetic or resistance)

   _____

   _____

3. What is the advantage of a sequencer over a contactor?

   _____

   _____

4. How many heaters can a typical package sequencer energize?

   _____

   _____

5. How is the fan motor started with a typical package sequencer?

   _____

   _____

6. What is the current-carrying capacity of each contact on the sequencer you worked with?

   _____

   _____

7. Do all electric furnaces use package sequencers?

   _____

   _____

8. If the number 2 contact failed to close on a package sequencer, would the number 3 contact close?

   _____

   _____

9. How should you determine the heat anticipator setting on the room thermostat when using a package sequencer?

   _____

   _____

10. How should you determine the heat anticipator setting on the room thermostat when using individual sequencers?

   _____

   _____

Instructor's Response

Name: _____ Date: _____

Job Sheet 4: Checking Electric Heating Elements Using an Ammeter

Upon completion of this job sheet, you should be able to check an electric heating system by using a clamp-on ammeter to make sure that all heating elements are drawing power.

PROCEDURES

1. Turn off the power to an electric furnace and remove the door to the control compartment.

2. Locate the conductor going to each electric heat element. Consult the diagram for the furnace you are working with. Make sure that you can clamp an ammeter around each conductor.

3. Turn the power on and listen for the sequencers to start each element. Wait for about 3 minutes; the fan should have started.

4. Clamp the ammeter on one conductor to each heating element and record the readings below. When the element is drawing current, it is heating.
   - Element number 1 _____ A   Element number 2 _____ A
   - Element number 3 _____ A   Element number 4 _____ A
   - Element number 5 _____ A   Element number 6 _____ A
   Note: Your furnace may have only two or three elements.

5. Turn the room thermostat off and, using the clamp-on ammeter, observe which elements are de-energized first.

6. Replace all panels with the correct fasteners.

QUESTIONS

1. What are the actual heating element wires made of?

   _____

   _____

2. If the resistance in a heater wire is reduced, what will the current do if the voltage is a constant?

   _____

   _____

3. How much heat would a 5-kW heater produce in Btuh?

   _____

   _____

4. What started the fan on the unit you were working on?

   _____

   _____

5. What is the difference between a contactor and a sequencer?

   _____

   _____

6. Why are sequencers preferred over contactors in duct work applications?

   _____

   _____

7. How much current would a 5-kW heater draw with 230 volts as the applied voltage?

   _____

   _____

8. What is the advantage of a package sequencer over a system with individual sequencers?

   _____

   _____

9. If the number one sequencer coil were to fail in a system with individual sequencers, would all the heat be off?

   _____

   _____

Instructor's Response

Name: _____   Date: _____

Job Sheet 5: Changing a Sequencer

Upon completion of this job sheet, you should be able to change a sequencer in an electric furnace.

PROCEDURES

1. Turn the power off and lock it out.

2. Check to make sure the power is off using the VOM.

3. Draw a wiring diagram showing all wires on the sequencer to be changed. If some of the wires are the same color, make a tag and mark them for easy replacement.

4. Remove the wires from the sequencer one at a time. Use care when removing wires from spade or push on terminals and don't pull the wire out of the connector. Use needle nose pliers. Check for properly crimped terminals.

5. Remove the sequencer to the outside of the unit. This would be just like having a new one.

6. Now, remount the sequencer.

7. Replace the wiring one at a time. Be very careful that each connection is tight. Electric heat pulls a lot of current and any loose connection will cause a problem.

8. When the sequencer is installed to your satisfaction, ask the instructor to approve the installation before you turn the power on.

9. Turn the power on and set the thermostat to call for heat.

10. After the unit has been on long enough to allow the sequencers to energize the heaters, use the ammeter to verify the heaters are operating.

11. Turn the unit off when you are satisfied it is working.

QUESTIONS

1. What would low voltage do to the capacity of an electric heater?

   _____

   _____

2. What would high voltage do to the capacity of an electric heater?

   _____

   _____

3. An electric heater has a resistance of 11.5 ohms and is operating at 230 volts. What would the current reading be for this heater? Show your work.

_____

_____

4. What would the power be for the above heater? Show your work.

_____

_____

5. What would the Btuh rating be on the above heater? Show your work.

_____

_____

6. If the voltage for the above heater were to be increased to 245 volts, what would the power and Btuh be? Show your work.

_____

_____

7. If the voltage were reduced to 208 volts, what would the power and Btuh be for the above unit? Show your work.

_____

_____

8. Using the above voltages, would the heater be operating within the + or − 10 percent recommended voltage?

_____

_____

9. What is the unit of measure the power company charges for electricity?

_____

_____

10. What starts the fan motor in the unit you worked on?

_____

_____

Instructor's Response

Name: _____   Date: _____

Job Sheet 6: Changing the Heating Element in an Electric Furnace

Upon completion of this job sheet, you should be able to change a heating element in an electric furnace using the correct techniques.

PROCEDURES

1. Turn the power off and lock it out, then check with VOM to be sure.

2. Remove the panel door to the electric heat section.

3. Check all wires to the heater to be changed for color coding, comparing them to the wiring diagram. If they are not coded, tag them so the correct wires can be placed back in their places. This is very important. You can't rely on memory.

4. Remove the wires, one at a time. If spade terminals are used, use care when removing the terminals. Needle nose pliers may be used by gripping only the connector. Do not pull on the wire.

5. When all the wires are clear, loosen the heater mounting and remove the heater. If there is any question about how the heater should be removed, consult your instructor.

6. Set the heater section on a work bench and record the following information.
   A. What is the resistance through the heating element? _____ ohms
   B. Does the element have a fuse link? (yes or no)

7. Now, replace the heating element in the furnace and fasten it.

8. Replace all wiring to the heating element. Make sure all connections are tight.

9. Have your instructor inspect the job.

10. When all is corrected, turn on the power and start the unit. Give it a few minutes to warm up and for all heaters to be energized.

11. Record the current and voltage of the heater.
    A. _____A   B. _____V   Compare with data on unit data plate.

12. Turn the unit off.

QUESTIONS

1. What is the heating element wire made of?

   _____

   _____

2. What is the purpose of the fuse link?

   _____

   _____

3. What other protection from overheating did this unit have?

_____

_____

4. How many elements did this furnace have?

_____

_____

5. If more than one, did it have stack sequencers, individual sequencers, or a package sequencer?

_____

_____

6. Using Ohm's law, what should the current for this heater have been? Show your work.

_____

_____

7. Are the actual current and the calculated current the same? If not, why not?

_____

_____

8. What is the Btuh for this heater? Show your work.

_____

_____

Instructor's Response

Name: _____   Date: _____

Job Sheet 7: Properties of Gases

Upon completion of this job sheet, you should be able to define the properties of natural, propane, and butane gases.

PROCEDURES

1. A cubic foot of air contains _____ percent nitrogen and _____ percent oxygen.

2. One thousand cubic feet of air contains _____ cubic feet of oxygen.

3. A gas burner with an output of 110,000 Btuh is proposed for a particular job. How much air will it take to support perfect combustion? _____ How much air will need to be planned for practical combustion?

4. If the above burner were to be used for propane, how much air would be required for perfect combustion? How much air will need to be planned for practical combustion? _____

5. A small building requires that the gas heat appliance have an output of 225,000 Btuh for heating. What furnace input would need to be furnished if the furnace has an operating efficiency of 80 percent? _____

6. What would be the fuel consumption for the above furnace in cubic feet per hour for natural gas? _____ What would the fuel consumption be for propane in cubic feet per hour? _____

7. What would be the fuel consumption for a natural gas furnace in the above problem if the furnace were to have an efficiency of 95 percent? _____

QUESTIONS

1. Air contains _____ percent oxygen and _____ percent nitrogen.

2. When natural gas is taken out of the ground, what must be done to prepare it to burn as a fuel?

   _____

   _____

3. Why are propane and butane called LP gases?

   _____

   _____

4. Which gas has the most carbon? (propane, butane, or natural gas) _____

5. What is the typical manifold pressure for natural gas? _____ LP gas? _____

6. State the specific gravity for the following gases.
   A. Natural gas _____
   B. Propane gas _____
   C. Butane gas _____

7. How many cubic feet per hour would a furnace that is 81 percent efficient consume while operating on propane? _____, butane? _____ , natural gas? _____

8. What is the ignition temperature for the following gases?
   A. Natural gas _____
   B. Propane gas _____
   C. Butane gas _____

9. Why is perfect combustion not used when setting the air adjustment for a gas burner?

   _____

   _____

10. Describe the difference in a power gas burner and an atmospheric gas burner.

   _____

   _____

Instructor's Response

Name: _____   Date: _____

Job Sheet 8: Gas Burners and Heat Exchangers

Upon completion of this job sheet, you should be able to recognize the differences in the various types of gas burners.

PROCEDURES

1. Proceed to a standard forced air gas furnace and turn off the gas, turn off the power, and lock it out.
   Note: This exercise is hard to perform on a furnace with force or induced draft configuration.

2. Remove the burner access panel.

3. Remove the flue pipe and draft diverter.

4. What type of burners are used with this furnace? (endshot, slotted port, drilled port, or ribbon)

   _____

5. How many burners are there? _____

6. What is the capacity for each burner? _____ Btuh

7. Using the flashlight, examine the heat exchanger and describe how the flue gases reach the draft diverter.

8. What material is the heat exchanger made of? _____

9. Replace all panels with the correct fasteners.

QUESTIONS

1. What is the purpose of the serpentine type of heat exchanger?

   _____

   _____

2. What is the purpose of the draft diverter?

   _____

   _____

3. If a heat exchanger were to rust and patches of rust were to fall on the burner, what could the results be?

_____

_____

4. What type of air adjustment did the burner have on the furnace you worked with?

_____

_____

5. The word "clamshell" applies to what component of a furnace?

_____

_____

6. Why should the combustion gases never be allowed to mix with the air from the conditioned space?

_____

_____

7. What causes a heat exchanger to become damaged?

_____

_____

8. What are the symptoms of a damaged heat exchanger?

_____

_____

9. How can a heat exchanger be examined for damage?

_____

_____

10. Where do the flue gases terminate?

_____

_____

Instructor's Response

Name: _____   Date: _____

Job Sheet 9: Gas Furnace Familiarization

Upon completion of this job sheet, you should be able to recognize and state the various components of a typical gas furnace.

PROCEDURES

1. With the power off, remove the front burner and blower compartment panels.

2. Fan information:

   Motor full-load amperage _____A   Type of motor _____
   Diameter of motor _____ in.   Shaft diameter _____ in.
   Motor rotation (looking at motor shaft) _____
   Fan wheel diameter _____ in.   Width _____ in.
   Number of motor speeds _____, high rpm _____, low rpm _____

3. Burner information:

   Type of burner _____   Number of burners _____
   Type of pilot safety _____   Gas valve voltage _____V
   Gas valve amperage _____A   Gas valve pipe size _____ in.

4. Unit nameplate information:

   Manufacturer _____   Model number _____
   Serial number _____   Type of gas _____
   Input capacity _____   Btu/h output capacity _____ Btu/h
   Voltage _____V   Recommended temperature rise _____ °F
   Control voltage _____V

5. Heat exchanger information:

   What is the heat exchanger made of? _____ (type of metal)
   Number of burner passages _____ Flu size _____ in.
   Type of heat exchanger? (upflow, downflow, or horizontal) _____

6. Replace all panels with the correct fasteners.

QUESTIONS:

1. What component transfers the heat from the products of combustion to the room air?

   _____

   _____

2. What is the typical gas manifold pressure for a natural gas furnace?

   _____

   _____

3. Which of the following gases requires a 100 percent gas shutoff? (natural or propane)

   _____

   _____

4. Why does this gas require a 100 percent shutoff?

   _____

   _____

5. What is the typical control voltage for a gas furnace?

   _____

   _____

6. Name two advantages for this particular control voltage.

   _____

   _____

7. What is the purpose of the drip leg in the gas piping located just before a gas appliance?

   _____

   _____

8. What is the typical line voltage for gas furnaces?

   _____

   _____

9. What is the purpose of the vent on a gas furnace?

   _____

   _____

10. What is the purpose of the draft diverter on a gas furnace?

   _____

   _____

Instructor's Response

Name: _____   Date: _____

Job Sheet 10: Identification of the Pilot Safety Feature of a Gas Furnace

Upon completion of this job sheet, you should be able to look at the controls of a typical gas furnace and determine the type of pilot safety features used.

PROCEDURES

1. Select a gas-burning furnace and shut the power off and lock it out.

2. Remove the panel to the burner compartment.

3. Remove the cover from the burner section if there is one.

4. Using the flashlight, examine the sensing element that is in the vicinity of the pilot light. Follow the sensing tube to its termination point to help identify it.

5. What type of pilot safety feature does this furnace have?

6. If the pilot light is lit, blow it out for the purpose of learning how to relight it.

7. Turn the power on and light the pilot light. You may find directions for the specific procedure in the furnace. If not, ask your instructor.

8. Turn the room thermostat to call for heat. Make sure that the burner ignites.

9. Allow the furnace to run until the fan starts, then turn the room thermostat to off and make sure the burner goes out.

10. Stand by until the fan stops running.

11. Replace all panels with the correct fasteners.

QUESTIONS

1. Why is it necessary to have a pilot safety shutoff on a gas furnace?

_____

_____

2. How long could gas enter a heat exchanger if a pilot light were to go out during burner operation with a thermocouple pilot safety?

_____

_____

3. Describe how a thermocouple works.

_____

_____

4. Describe a mercury sensor application.

_____

_____

5. What is the typical heat content of a cubic foot of natural gas?

_____

_____

6. Where does natural gas come from?

_____

_____

7. What is the typical pressure for a natural gas furnace manifold?

_____

_____

8. What component reduces the main pressure for a typical gas furnace?

_____

_____

9. How does the gas company make the consumer aware of a gas leak?

_____

_____

10. How does the gas company charge the consumer for gas consumption?

_____

_____

Instructor's Response

Name: _____   Date: _____

Job Sheet 11: Changing a Gas Valve on a Gas Furnace

Upon completion of this job sheet, you should be able to change a gas valve on a typical gas furnace, using the correct tools and procedures.

PROCEDURES

1. Turn the power off and lock it out, check it with the VOM.

2. Turn the gas off at the main valve before the unit to be serviced.

3. Remove the door from the furnace. Watch how it removes, so you will know how to replace it.

4. Make a wiring diagram of the wiring to be removed. The wiring may need to be tagged for proper identification. Remove the wires from the gas valve. Use needle nose pliers on the connectors if they are spade type. Do not pull the wires out of their connectors.

5. Look at the gas piping and decide where to take it apart. There may be a pipe union or a flare nut connection to the inlet of the system. Either one may be worked with.

6. Remove the pilot line connections from the gas valve. Use the correct size end wrench and do not damage the fittings.

7. Use the adjustable wrenches for flare fitting connections and pipe wrenches for gas piping connections and disassemble the piping to the gas valve. Look for square shoulders on the gas valve and be sure to keep the wrenches on the same side of the gas valve. Too much pressure can easily be applied to a gas valve by having one wrench on one side of the valve and the other on the piping on the other side of the valve.

8. Remove the valve from the gas manifold. Use care not to stress the valve or the manifold.

9. While you have the valve out of the system, examine the valve for some of its features, such as where a thermocouple or spark igniter connects. Look for damage.

10. After completely removing the gas valve, treat it like a new one and fasten it back to the gas manifold. Be sure to use thread seal on all external pipe threads. Do not use excessive amounts of thread seal. Do not use thread seal on flare fittings.

11. Fasten the inlet piping back to the gas valve.

12. Fasten the pilot line back using the correct wrench.

13. Fasten all wiring back using the wiring diagram.

14. Ask your instructor to look at the job and approve, then turn the power and gas on. Immediately use soap bubbles and leak check the gas line up to the gas valve.

15. Turn the thermostat to call for heat. When the burner lights, you have gas to the valve outlet and the pilot light. Leak check immediately.

16. When you are satisfied there are no leaks, turn off the furnace.

17. Use a wet cloth and clean any soap bubble residue off the fittings.

QUESTIONS

1. Did the gas valve on this system have a square shoulder for holding with a wrench?

_____

_____

2. Did you notice a gas leak after the repair?

_____

_____

3. Why is thread seal used on threaded pipe fittings?

_____

_____

4. Why is thread seal not used on flared fittings?

_____

_____

Instructor's Response

Chapter 11 — Appliance Repair and Replacement

Upon completion of this chapter, you will be able to:

Skill-Based

- ⊗ Replace a gas stove.
- ⊗ Replace an electric stove.
- ⊗ Replace a heating element on an electric stove.
- ⊗ Replace an oven heating element.
- ⊗ Repair a range hood.
- ⊗ Replace a range hood.

Keywords

Heating Element

   a device used to transform electricity into heat through electrical resistance.

Introduction

Before attempting to replace any electrical appliance it is recommended that the facility mainte-nance personnel first review Chapter 5, "Basic Electric Theory." Always take the proper precautions when working with electricity. Also, be sure to check all state and local building codes when repair-ing any gas or electric appliances.

Chapter Review Questions and Exercises

SHORT ANSWER

1. List the steps for repairing an oven that will not heat properly.

   _____

   _____

   _____

2. List the steps for repairing a gas burner that will not light.

   _____

   _____

   _____

3. List the steps for replacing a heating element on an electric stove.

_____

_____

_____

4. List the steps for replacing a dishwasher.

_____

_____

_____

5. List the steps for unclogging an exhaust fan.

_____

_____

_____

6. What should a technician check if a range does not heat properly?

_____

_____

_____

7. What should a technician check if the surface burner is getting hot hot?

_____

_____

_____

8. What should a technician check if a dishwasher motor does not start but the motor hums?

_____

_____

_____

9. What should a technician check if a dishwasher does not dry properly?

_____

_____

_____

10. When is it necessary to replace an appliance?

_____

_____

_____

Name: _____   Date: _____

Job Sheet 1: Checklist for Replacing a Washing Machine

- The checklist below is provided as a quick reference for replacing a washing machine. It is recommended that when replacing a washing machine, you should save a copy of this checklist with the washing machine documentation for future reference.

Step	Description	Complete	Comment
1.	Turn off power to the old washer.	❏	
2.	Remove the old washer.	❏	
3.	Clean and dry the floor and move the new washer close to the connection point.	❏	
4.	Fasten the drain hose to the washer with a hose clamp. Avoid tightening it too much or you might strip the screw.	❏	
5.	Attach the water hoses to the washer. The hot and cold on the taps and on the washer are usually clearly marked. Red indicates hot; blue indicates cold.	❏	
6.	Plug the washing machine in and move it almost into position, placing the drain hose in the drainpipe where you can reach it.	❏	
7.	Push the washer the rest of the way into position, being careful not to crimp the hoses.	❏	
8.	Leave about an inch and a half of space around the washer to allow room for it to vibrate.	❏	
9.	Turn the water faucets on.	❏	
10.	Turn the power back on.	❏	
11.	Run a cycle without clothes or detergent before you use the machine to clear the water pipes and make sure the drainage is adequate.	❏	

Date Completed: _____

Completed by: _____

Instructor's Response

Name: _____   Date: _____

Job Sheet 2: Checklist for Replacing a Range Hood

- The checklist below is provided as a quick reference for replacing a range hood. It is recommended that when replacing a range hood, you should save a copy of this checklist with the range hood documentation for future reference.

Step	Description	Complete	Comment
1.	Remove the old range hood.	☐	
2.	On the new range hood, remove the filter, fan, and electrical housing cover. Remove the knockouts for the electrical cable and the duct.	☐	
3.	Protect the surface of the cooktop with heavy cardboard and set the range hood on top of it. Then connect the house wiring to the hood. Connect the house black wire to the hood black wire and the house white wire to the hood white wire. Then connect the house ground wire under the ground screw and tighten the cable clamp onto the house wiring.	☐	
4.	Using the mounting screws to install the hood, slide the hood toward the wall until the mounting screws are engaged. Tighten the screws securely with a long-handled screwdriver. Then replace the bottom cover.	☐	
5.	Fasten the ductwork to the hood by using duct tape to secure joints and make them airtight.	☐	
6.	Install the light bulbs, replace the filters, turn on the power at the service panel, and check for proper operation.	☐	

Date Completed: _____

Completed by: _____

Instructor's Response

Chapter 12   Trash Compactors

OBJECTIVES

Upon completion of this chapter, you will be able to:

Knowledge-Based

- Explain the purpose of interlock safety device.

Skill-Based

- Perform general maintenance procedures.
- Perform general maintenance of hydraulic devices.
- Perform a test of interlock safety device.
- Check the general condition of dumpsters.

Keywords

Dumpsters

a large waste receptacle

Introduction

When you work with or around hydraulics, certain precautions should be followed to ensure your safety.

- *Always wear safety glasses.*
- *Always replace oil to manufacturers' specifications.*
- *Never over-tighten fittings.*
- *Never check for leaks by using your hands.*
- *Never substitute parts on a hydraulic system.*
- *If fittings, hoses, or valves show signs of wear, replace at once.*
- *Treat hydraulic systems with the same respect as electrical systems.*

Although hydraulic systems do not pose the same threat as electrical systems, they are not without danger. Keep in mind that a hydraulic system works by compressing a fluid to accomplish work. Unlike with pneumatic systems (air), the fluid in a hydraulic system has very little play and therefore tends to produce higher working pressures than pneumatic systems.

Chapter Review Questions and Exercises

SHORT ANSWER

1. If a trash compactor will not operate, what should a technician check?

_____

_____

_____

2. What should the technician check if the compactor does not complete its cycle?

_____

_____

_____

3. What should the technician check if the drawer will not open on a trash compactor?

_____

_____

_____

4. Why is it important to clean and deodorize a trash compactor?

_____

_____

_____

5. What is the purpose of the interlocks on a trash compactor?

_____

_____

_____

6. List the steps for cleaning a trash compactor.

_____

_____

_____

7. What should the technician check if the motor on a trash compactor runs but the trash does not compact?

_____

_____

_____

Name: _____   Date: _____

Job Sheet 1: Cleaning a Trash Compactor

• The checklist below is provided as a quick reference for cleaning a trash compactor. It is recommended that when cleaning a trash compactor you save a copy of this checklist with the trash compactor documentation for future reference.

Note: Always wear thick, sturdy gloves when cleaning your compactor.

Step	Description	Complete	Comment
1.	Unplug the compactor.	☐	
2.	Remove the bag and caddy, or bin, and follow the manufacturer's cleaning instructions.	☐	
3.	Vacuum the inside.	☐	
4.	Clean inside and outside of the compactor with warm, soapy water. Rinse and dry.	☐	
5.	Close the drawer and replace the caddy with a new bag.	☐	

Note: Periodically, check and replace the air freshener or charcoal filter.

Date Completed: _____

Completed by: _____

Instructor's Response

Chapter 13   Elevators

Upon completion of this chapter, you will be able to:

Skill-Based

- ✪ Check and inspect floor leveling.
- ✪ Check operation of elevators.
- ✪ Perform a test on elevator doors.

Keyword

Elevator platform

Introduction

Elevator safety is typically controlled by state and local building codes; therefore, all maintenance should be done by a certified elevator mechanic. Never attempt to adjust an elevator on your own. Despite this limitation, a few items can be checked by a facility maintenance technician.

Chapter Review Questions and Exercises

SHORT ANSWER

1. List the steps for checking the reaction time of an elevator door.

_____

_____

_____

2. Using the Internet, research the local and state codes governing elevators in your area.

_____

_____

_____

3. What is the minimum door width for an elevator to be considered ADA compliant?

_____

_____

_____

4. What is the minimum width and depth for an elevator car to be considered ADA compliant?

_____

_____

_____

5. For an elevator to be considered ADA compliant the elevator call button must be _____ inches from the floor.

Name: _____   Date: _____

Job Sheet 1: Elevator Checklist

* The checklist below is provided as a quick reference for elevator maintenance.
* Elevators and other lifting devices are inspected regularly, per local code, and a preventive maintenance contract is established.

_____ Completed   _____ in Progress   _____ Not Planned

Date Completed: _____

Completed by: _____

Instructor's Response

Chapter 14 Pest Prevention and Control

Upon completion of this chapter, you will be able to:

Knowledge-Based

⊗ Follow applicable safety procedures.

Skill-Based

⊗ Recognize the sources of damage caused by pests.

⊗ Select and apply proper techniques, chemicals, and/or materials to eradicate and/or prevent pest infiltration.

Keywords

Insecticide

Pesticide

Introduction

Before working with pesticides always read and follow manufacturers' directions. Never mix pesticides. Mixing certain chemicals can produce toxic gas. In addition, mixing pesticides will change the chemical properties of the pesticides and can result in a toxic mixture that if released into the environment could cause adverse effects to humans, plants, and animals.

Chapter Review Questions and Exercises

TRUE/FALSE

1. _____ The pest control method used for ants is identical for cockroaches. Explain your answer.

2. _____ A motorized sprayer should never be used when applying pesticides. Explain your answer.

3. _____ If the first application of a pesticide does not work, then it is okay to mix boric acid with the pesticide and reapply.

SHORT ANSWER

4. What is the difference between a house fly and a fruit fly?

_____

_____

5. How are rodents controlled?

_____

_____

6. What is the first step to preventing and controlling pests?

_____

_____

7. What can a facility maintenance technician do to prevent termites?

_____

_____

8. Why should pesticides be stored only in their original containers with labels visible and intact?

_____

_____

9. When examining an area to be treated with a pesticide, what should a technician do if something is found that could be harmed by the pesticide?

_____

_____

10. Why is it important not to mix pesticides?

_____

_____

11. List five nonpesticidal methods for controlling pests.

_____

_____

12. Why should a facility maintenance technician not kill a spider?

_____

_____

Name: _____   Date: _____

Job Sheet 1: Pest Control

• The checklist below is provided as a quick reference for pest control.

Step	Description		Complete	Comment
1.	Identify the pests.		☐	_____
2.	Recognize the source.		☐	_____
2.1.	Check exterior doors. If you can see light under the door, this is a potential problem.		☐	_____
2.1.1.		Install door thresholds.	☐	_____
2.1.2.		Seal around windows.	☐	_____
2.2.	Check for windows that don't fit properly or have holes in the screens.		☐	_____
2.2.1.		Seal around windows.	☐	_____
2.2.2.		Use mesh or screens to fix holes in the screen.	☐	_____
2.3.	Check for openings around any objects that penetrate the building's foundation, such as plumbing, electrical service, telephone wires, HVAC, and so on.		☐	_____
2.3.1.		Seal around these objects.	☐	_____
2.3.2.		Remove materials against the foundation of a building.	☐	_____
2.3.3.		Install motion sensors on outdoor lights.	☐	_____
2.3.4.		Ensure a plant-free zone of about 12 inches around the building.	☐	_____
2.3.5.		Place outdoor garbage containers away from the buildings and on concrete or asphalt slabs	☐	_____
2.3.6.		Ensure that surrounding areas are clean.	☐	_____

Pest Are Eliminated	
3.	Identify the pests.
4.	Become familiar with methods of control.
5.	Estimate level of infestation.
6.	Determine method of application.
7.	Select pesticide for best control and least hazard.

Date Completed: _____

Completed by: _____

Instructor's Response

Chapter 15 Landscaping and Groundskeeping

OBJECTIVES

Upon completion of this chapter, you will be able to:

Skill-Based

- ☒ Maintain and police grounds including mowing, edging, planting, mulching, leaf removal, and other assigned tasks.
- ☒ Perform basic small-engine repair and preventive maintenance according to manufacturer's specifications.
- ☒ Perform basic swimming pool maintenance that does not require certification.
- ☒ Remove refuse and snow as required.
- ☒ Maintain public areas including hallways, kitchens, and lobbies.
- ☒ Repair asphalt by using cold-patch material.

Keywords

Groundskeeping	Organic mulch
Landscaper	Inorganic mulch
Belly deck	

Introduction

Effective groundskeeping is much more than just mowing the grass and picking up a little trash. It requires the removal of ice and snow, planting and maintaining greenery, and more. In short, it involves the general upkeep of the building grounds. Groundskeeping should not be confused with landscaping. Landscaping, on the other hand, is an activity in which a person modifies the feature of a property (such as constructing retaining walls, plants shrubs, etc.). Often it is the facility maintenance person who is responsible for making minor modifications to the property. Therefore in addition to being able to keeping the grounds a facility maintenance technician should have a good understanding of the principles and techniques of landscaping.

Landscaping Resources

Because it is almost impossible for a facility maintenance technician to have a working knowledge of all the different landscaping techniques and principles, the technician should have a good understanding of how to research the different techniques using the Internet. For the most part the Internet can be an extremely useful tool for doing research; however, the technician should exercise care when selecting a Web site to trust.

Any sites local, state, and federal government sites as well as landscapling suppliers and manufactures can be trusted that the information is correct. For additional information on plants the greenhouse or nursery in which the plants were purchased can be trusted. Finally, professional organizations Web sites can be trusted.

Chapter Review Questions and Exercises

TRUE/FALSE

1. _____ A string trimmer should never be used to edge around concrete and pavement. Explain your answer.

2. _____ Because irrigation systems are placed underground, there is no need to winterize.

SHORT ANSWER

3. How often should grass be mowed?

_____

_____

4. What is mulch?

_____

_____

5. List the steps for removing snow.

_____

_____

6. How should a large hole be repaired using cold patch?

_____

_____

7. What are the steps for planting a shrub?

_____

_____

8. What is the purpose of aeration?

_____

_____

9. How short should *Zoysia matrella* grass be cut?

_____

_____

10. What is the purpose of edging?

_____

_____

Name: _____   Date: _____

Job Sheet 1: Maintaining Public Areas

- The checklist below is provided as a quick reference for maintaining public areas.

Description	Complete	Comment
Bathroom		
Disinfected daily.	☐	_____
Paper products restocked daily.	☐	_____
Thoroughly cleaned once a week.	☐	_____
Hand soap replaced as needed.	☐	_____
Stairways		
Thoroughly cleaned at least once a week.	☐	_____
Swept daily.	☐	_____
Hallway		
Buffed to a shine at least once a week, twice if time permits.	☐	_____
General cleaning daily.	☐	_____
Walls cleaned weekly and spot cleaned as needed.	☐	_____
Alcoves and shelves thoroughly cleaned.	☐	_____
Water fountains disinfected daily.	☐	_____
Trash Cans		
Emptied daily.	☐	_____
Disinfected and thoroughly cleaned once a week.	☐	_____
Lobby		
Thoroughly cleaned weekly.	☐	_____
General cleaning daily.	☐	_____
Porches, Patios, and Dumpster Pads		
Broom swept as needed.	☐	_____

Date Completed: _____

Completed by: _____

Instructor's Response

Name: _____   Date: _____

Job Sheet 2: Maintaining a Small Engine

- The checklist below is provided as a quick reference for maintaining a small engine.

5 Hours	25 Hours	50 Hours	100 Hours
Check oil level.	_____	Change oil.	_____
			_____
	Replace oil-foam element.	_____	
	Replace air-cleaner cartridge.		
		_____	
			Clean cooling system. Replace spark plug(s).
		Inspect spark-arrester.	_____
		Replace in-line fuel filter.	_____
Check the blade and engine mounting fasteners, making sure they are all tight.			
	_____	_____	_____
Clean built-up grass clippings and dirt.	_____	_____	_____
Clean grass clippings and debris under belt cover and drive belt (self-propelled models).	_____	_____	_____
		Sharpen or replace blade.	

Date Completed: _____

Completed by: _____

Instructor's Response

Name: _____   Date: _____

Job Sheet 3: Plant Research Using the Internet

Plant Research Using the Internet

Upon completion of this job sheet, you will be able to research growing conditions of various grasses using the Internet.

Procedure

Using the Internet determine the optimum growing conditions for the following grasses.

Centipede
Common Bermuda
Hybrid Bermuda
Kentucky Blue
St. Augustine
Zoysia japonica
Zoysia matrella

Instructor's Response

Name: _____   Date: _____

Job Sheet 4: Researching Swimming Pool Cleaning Methods

Swimming Pool Cleaning Methods

Upon completion of this job sheet, you should be able to use the Internet to determine the types and amount of chemicals needed to clean and treat the water in a swimming pool.

Procedure

Using the Internet, determine the best way in which to clean and treat the water in an average-sized public swimming pool.

Instructor's Response

Chapter 16 Basic Math for Facility Maintenance Technicians

By the end of this chapter, you will be able to:

Knowledge-Based

- State the difference between a real number and a whole number.
- State the difference between an integer and a whole number.

Skill-Based

- Add whole and real numbers and fractions.
- Subtract whole and real numbers and fractions.
- Multiply whole and real numbers and fractions.
- Divide whole and real numbers and fractions.
- Solve problems involving multiple operations with whole numbers.

Keywords

Integer

Real number

Whole numbers

Fraction

Introduction

Having a firm grasp of the principles of basic math is essential for the facility maintenance technician. Without this knowledge, calculating the amount of materials needed to complete a repair and/or installation would be difficult if not impossible. In addition technicians wishing to advance into management will not be able to do so without the understanding of basic math.

Chapter Review Questions and Exercises

1. 0.91 as a percent is _____.

2. 8.731 as a percent is _____.

3. $\frac{11}{16}$ as a percent is _____.

4. 60 is 80 percent of _____.

5. 25% of 120 is _____.

6. 173.8% of 87.20 is _____. (round to 1 decimal place)

7. 519 is 120% of _____.

8. A casting, when first poured, is 14.872 inches long. The casting shrinks 0.098 inch when it cools. The percent of shrinkage is _____. (round to 2 decimal places)

9. A carpenter completes a home remodeling job in 85% of the estimated time. The estimated time is 136 hours. The job actually takes _____ hours to complete. (round to the nearest hour)

10. An engine loses 6.70 horsepower through friction. The power loss is 5.20% of the total rated horsepower. The total horsepower rating is _____. (round to the nearest whole horsepower)

11. An alloy of brass is composed of 84.6% copper, 5.20% tin, 6.70% lead, and zinc 3.5%. The number of pounds of zinc required to make 520.0 pounds of alloy is _____.

12. The first shift of a manufacturing firm produces 3.6% defective pieces out of a total production of 2,555 pieces. The second shift produces 5.8% defective pieces out of a total production of 2,072 pieces. The first shift produces _____ more acceptable pieces than the second shift.

Name: _____   Date: _____

Job Sheet 1: Percentages

Percentages

Upon completion of this job sheet, you should be work with percentages.

Procedure

1. If the list price of a product is 658.50 and it is discounted at 57% what is its final price?
2. If the list price of a product is 255.84 and it is discounted at 65% what is its final price?
3. If the list price of a product is 465.42 and it is discounted at 29% what is its final price?
4. If the list price of a product is 554.92 and it is discounted at 75% what is its final price?
5. If the list price of a product is 422.70 and it is discounted at 42% what is its final price?
6. If the list price of a product is 435.97 and it is discounted at 18% what is its final price?
7. If the list price of a product is 696.37 and it is discounted at 13% what is its final price?
8. If the list price of a product is 367.84 and it is discounted at 64% what is its final price?
9. If the list price of a product is 614.33 and it is discounted at 94% what is its final price?
10. If the list price of a product is 678.42 and it is discounted at 75% what is its final price?
11. If the list price of a product is 255.21 and it is discounted at 3% what is its final price?
12. If the list price of a product is 977.49 and it is discounted at 86% what is its final price?
13. If the list price of a product is 212.48 and it is discounted at 16% what is its final price?
14. If the list price of a product is 61.81 and it is discounted at 48% what is its final price?
15. If the list price of a product is 954.99 and it is discounted at 7% what is its final price?

Instructor's Response

Name: _____   Date: _____

Job Sheet 2: Angles

Percentages

Upon completion of this job sheet, you should be able to work with angles.

Procedure

1. Determine the following angle.

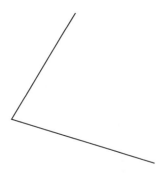

2. Determine the following angle.

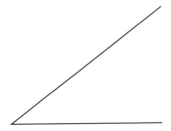

3. Determine the following angle.

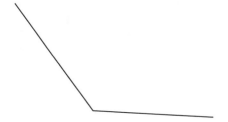

4. Determine the following angle.

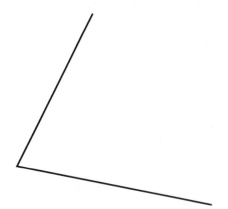

5. Determine the following angle.

Instructor's Response

Name: _____   Date: _____

Job Sheet 3: Area and Volume I

Area

Upon completion of this job sheet, you should be able to calculate area and volume.

Procedure

1. The volume of a prism with a height 4.80 inches and a base area 150.0 square inches is

   _____.

2. An excavation for a building foundation is 55.0 feet long, 36.5 feet wide, and 12.00 feet deep.
   _____ cubic yards of soil are removed.

3. An excavation for a building foundation is 55.0 feet long, 36.5 feet wide, and 12.00 feet deep.
   _____ truck loads are required to haul the soil from the building site if the average truck
   load is $5\frac{1}{2}$ cubic yards.

4. The height of a prism with a base area of 0.90 square meter and a volume of 1.62 meters is

   _____.

5. A rectangular shipping crate is 3 feet wide, 5 feet long, and 2'6" high. The total surface area of the crate
   is _____.

6. A rectangular carton is designed to have a volume of 2.05 cubic meters. The carton is 1.36 meters high
   and 1.75 meters long. The width of the carton is _____.

7. The volume of a regular pyramid which has a height of 3.40 feet and a base area of 5.70 square feet is

   _____.

8. A regular pyramid with a base area of 60.0 square inches contains 274 cubic inches of material. The height
   of the pyramid is _____.

9. A regular pyramid has a base perimeter of 4'6.0" and a slant height of 9.00 inches. The lateral area of the
   pyramid is _____.

10. The base area of a wooden form in the shape of a regular pyramid is 3.26 square meters. The form contains
    2.60 cubic meters of airspace. The form is _____ meters high.

Refer to the length of angle iron shown for the following problems. Round the answer to two significant digits.

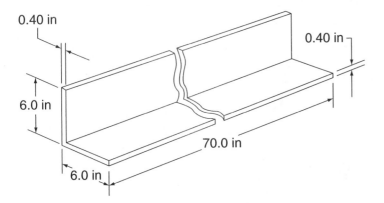

0.40 in

0.40 in

6.0 in

70.0 in

6.0 in

11. The volume of the angle iron is _____ .

12. If the material weighs 490 pounds per cubic foot, the weight of the angle iron is _____ .

A wooden planter is shown in the shape of a frustum of a pyramid with square bases. Refer to the figure for the following problems. Disregard the thickness of the wood.

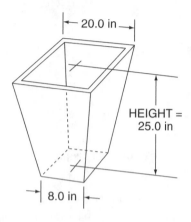

13. _____ cubic feet of soil are contained in the planter when full.

Instructor's Response

Name: _____   Date: _____

Job Sheet 4: Area and Volume II

Area

Upon completion of this job sheet, you should be able to calculate area and volume.

Procedure

1. The volume of a sphere with a diameter of 5.20 inches is _____.

2. A force of 28,200 pounds pulls on a steel cylinder that has a diameter 0.920 inches. The force pulling on 1 square inch of cross-sectional area is _____.

3. The sides and top of a cylindrical fuel storage tank are painted with two coats. The tank is 76.8 feet in diameter and 49.5 feet high. If 1 gallon of paint covers 550 square feet, _____ gallons are required. (round to nearest gallon)

4. A support column in the shape of a frustum of a right circular cone has a slant height of 12.25 feet. The smaller base is 13.00 inches in diameter and the larger base is 18.75 inches in diameter. The lateral area in square feet of the column is _____.

5. _____ gallons can be held by a container in the shape of a hemisphere with a 1'6.0" diameter.

6. A container with an open top is shown. The capacity of the container in liters is _____.

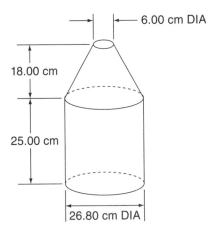

- 6.00 cm DIA
- 18.00 cm
- 25.00 cm
- 26.80 cm DIA

Instructor's Response

Chapter 17

Blueprint Reading for Facility Maintenance Technicians

Upon completion of this unit, you should be able to

Knowledge-Based

- ⊗ Identify the various views of an orthographic drawing.
- ⊗ Identify the various symbols used on plumbing plans to represent piping types, fittings, and symbols on a plumbing plan.
- ⊗ Identify the various symbols used on HVAC plans to represent HVAC line types, ducts, and equipment.
- ⊗ Identify the various symbols used on electrical plans to represent wiring, switches, fixtures, and so on.
- ⊗ Understand standard abbreviations and symbols used on blueprints.

Skill-Based

- ⊗ Determine the length of objects presented on a blueprint using an architect's scale and/or tape measure.
- ⊗ Determine the angle of a line on a blueprint using a protractor.
- ⊗ Understand standard abbreviations and symbols used on a blueprint.

Keywords

Linear measurement

English system of measurement

Metric system of measurement

Introduction

The basic function of facility maintenance technicians is to maintain the day-to-day operations of the equipment and facility in which they are responsible for. This may entail changing a lighting fixture to lawn maintenance. For some areas of this occupation, technical illustration and documentation is not necessary in completing a particular task. However, in other areas it is critical that the facility maintenance person have a good understanding of basic blueprint reading in order to troubleshoot, repair, and/or maintain equipment. For example, when repairing an appliance, an electrical schematic is necessary to properly trace the electrical connections.

Blueprint Reading Guidelines

When reading blueprints several key topics must be taken into consideration, they are:

- Linear measurement is defined as the measurement of two points along a straight line.
- All objects, whether they are man-made or the result of natural conditions and/or forces, consist of points and lines.
- When a mechanical drawing is created on the computer, typically it is drawn at a scale of 1:1 (also known as full scale). It is the responsibilities of the draftsman to note the scale at which the drawing was created.
- The title block region is typically located along the bottom edge or lower right-hand corner of the drawing.
- The length and position of items not dimensioned on a drawing can be determined using a ruler.
- Angles are typically given on a blueprint with either a dimension or a leader.
- Standards abbreviations and symbols have been developed for the engineering and architectural community that not only facilitate the development of blueprints but also ensure consistency in their interpretation.
- The primary view used on an engineering drawing are front view, right side view, left side view, top view, bottom view, and rear view.
- Architectural drawings are typically not labeled the same as engineering drawings; however, in reality they are still created based on the same principles.
- The floor plan provides a representation of where to locate the major items of a home.
- The plan view shows the location of walls, doors, windows, cabinets, appliances, and plumbing fixtures.
- The plumbing plan contains the size and location of each piping system contained in the proposed structure.
- The HVAC plan contains the size and location of each HVAC system contained in the proposed structure.
- An elevation is an orthographic drawing that shows one side of a building.

Chapter Review Questions and Exercises

TRUE/FALSE

1. When a drawing is created, one of the responsibilities of the drafter is to note the scale at which the drawing was created.
2. True/False; Standards abbreviations and symbols have been developed for the engineering and architectural community that not only facilitate the development of blueprints but also ensure consistency in their interpretation.

COMPLETION

3. The _____ scale is a type of ruler in which a range of precalibrated ratios are illustrated.
4. In a right triangle, there is a direct relationship between _____.
5. The _____ function is defined as the ratio of the side opposite to an acute angle divided by the hypotenuse.
6. The _____ function is defined as the ratio of the side adjacent to an acute angle divided by the hypotenuse.
7. The _____ function is defined as the ratio of the side opposite to an acute angle divided by the side adjacent.
8. _____ drawings are typically not labeled the same as engineering drawings; however, in reality they are still created based on the same principles.
9. The _____ plan provides a representation of where to locate the major items of a home.
10. What information is shown on a floor plan?
11. What information is shown on a plumbing plan?
12. What information is shown on a HVAC plan?

Name: _____   Date: _____

Job Sheet 1: Reading a Scale

Reading a Scale

Upon completion of this job sheet, you should be able to read a scale.

Procedure

1. In the space at right record the measurements of the dimension lines.

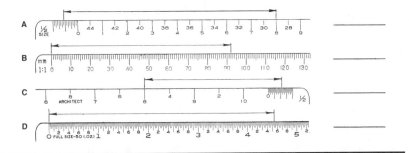

2. Add dimension lines to represent the measures that are listed at the right.

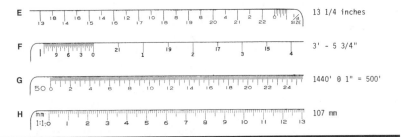

E	13 1/4 inches
F	3' - 5 3/4"
G	1440' @ 1" = 500'
H	107 mm

3. Measure from the left end of the long line, and mark the distance with a sharp, dark, vertical line. Use the distance and scale given at the right of each line.

I	2.77	decimal inch
J	102.5	metric 1:1
K	14 13/16	1/4 size
L	0'-11 1/2"	1/4" = 1'-0"
M	47.0 mm	metric 1:2
N	6 1/2"	half size
O	205.5 ft.	1" = 50'-0"

Instructor's Response

Name: _____   Date: _____

Job Sheet 2: Identifying Line types

Identifying Line types

Upon completion of this job sheet, you should be able to identify different line types on a drawing.

Procedure

Identify the different types of lines used on engineering and architectural drawings.

A. ― ― ― ― ― ― ― ― ― ― ― ―

B. ── ─ ── ── ─ ── ── ─ ──

C. ────────────────

D. ── ── ── ── ── ── ──

E. ▬ ▬ ▬ ▬ ▬ ▬ ▬ ▬ ▬

F. ▬▬▬▬▬▬▬▬▬▬▬▬

Instructor's Response

Name: _____   Date: _____

Job Sheet 3: Determining Angles

Determine Angles

Upon completion of this job sheet, you should be able to identify angles.

Procedure

Determine the angles of the following:

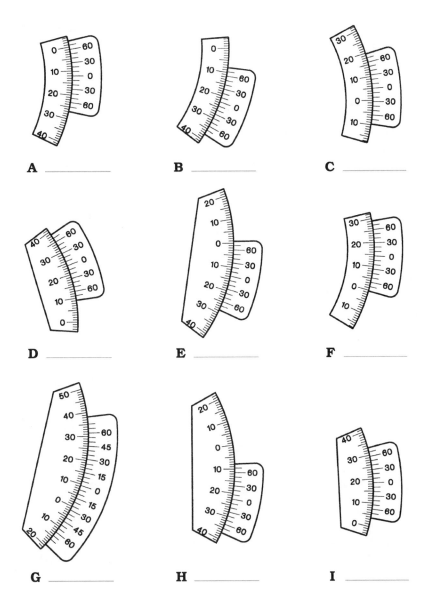

A _____   B _____   C _____

D _____   E _____   F _____

G _____   H _____   I _____

Instructor's Response